YOUR
BEST
SHOT

THE PERSONALIZED SYSTEM FOR OPTIMAL
WEIGHT HEALTH—GLP-1 SHOT OR NOT

Ashley Koff, RD

HARPERONE
An Imprint of HarperCollinsPublishers

CONTENTS

WELCOME TO YOUR BEST SHOT

He wasn't my first patient. But he remains one of the more memorable.

He was metabolically unhealthy—diabetes, obesity, elevated cholesterol, and high blood pressure. After all sorts of medications and diets, the lone remaining option to save his life was weight-loss surgery. The catch: To qualify, he needed to lose fifty pounds. His doctors had written him off. They called me as a last resort.

I'll share this patient's full story later, but suffice to say, he lost fifty-one pounds within six weeks and got the okay for the surgery. His journey shows that success in weight health could happen then, in the pre-GLP-1-shot era, as it can now—with or without the Shot.

In my practice over twenty-plus years, working with hundreds of patients, I've collected and developed a shed's worth of tools. The Shot is now one of those tools. For some, it's a game changer; for others, it's not effective or available; for most, it's not necessary. That's the benefit of a robust tool kit—lots of options; you discover what works for you today.

Here are a few other patient stories. Together, we developed a fully personalized plan to get to weight health on *their* terms—as we're here to do for you.

Tammy sought me out when she'd been on the Shot for three

months. She had lost fifteen pounds. Then she stopped losing. I quickly assessed that those fifteen pounds she'd lost included a lot of hard-earned lean body mass. She was concerned that she was doomed to be a Shot failure.

Stacy, a forty-year-old mom of three, was feeling awful, inside and out. She'd tried many diets, without the results she was after. The Shot, she was clear, wasn't a tool for her. She wasn't comfortable with the whole idea, and it didn't fit her budget. She came to me to find out whether there was something she could do to gain similar benefits. (Spoiler alert: Yes.)

Into her forties, Jan had managed her weight health proactively and successfully. Then her life turned into a series of car wrecks. The first was literal, resulting in back injuries, back pain, and extended treatments. The second was an out-of-nowhere divorce. The third was the financial crash that resulted. A major side effect of the three was forty pounds of fat around her middle, prediabetic blood sugar, and elevated inflammation. Six months on a semaglutide helped her shed thirty pounds and get her numbers back within a healthy range, as well as knock out the residual back pain. She told me, "I'm ready to come off this. But I don't want to be one of those statistics where they gain it all back and then some."

And then there's me. I became a practitioner because of my checkered experiences as a patient. Looking for what was better for me and—through trial and error, and significant work—finding it enabled me to build a system that can deliver better for others, too.

We'll do deeper dives into these stories and others to explore options, nuances, the impact of choices, and some cautionary tales. My story will be as accurate as I can recall. My patients' names are pseudonyms, but the methods and results are real.

My aim in sharing these stories isn't the familiar "testimonial" before/after, "I lost seven hundred pounds in ten days and looked stunning for my niece's wedding on the X plan." As different as my patients' stories are, they all have one common plotline:

Our bodies are a weight-health **ecosystem**. When optimally resourced, they run better. To regulate the ecosystem, there's a set of weight-health hormones. Let's meet the Switch.

YOUR BEST SHOT IS A PERSONALIZED SYSTEM

The system you'll build and personalize is the opposite of one-stop shopping, so it will have multiple moving parts—because you have multiple moving parts. You should not have to use time and brain space to identify and sort what you need.

Via this QR Code or link (https://thebetternutritionprogram .com/your-best-shot) you've got access to a whole range of resources that help you put your plan's components together and keep track of them, including live human help and coaching. As you'll learn in the upcoming stories, optimization is an interactive and ongoing process.

Through each part of the book, these resources and the Operator's Manual help you take action (or not) and track your plan's progression and results.

PART I

SWITCH STORIES

CHAPTER 1

THE SHOT STORY

THE WORLD BEFORE THE SHOT

The body has different systems.

Each system has a doctor, a prescription, and a nutrition plan.

Tens of millions of Americans are battling health challenges.

For many, the diagnoses include "obese" and "diabetic."

Recommendations say: Go on a diet. Eat until you are 80 percent full.

If your blood sugar is elevated, cut out carbs and added sugar.

Some people are insulin resistant or produce too little insulin; there are medications to address these.

Ninety-three percent of Americans are metabolically unhealthy.[1]

The rest of the industrialized world is catching up fast.

Current medications and diets aren't addressing the crisis.

THE SHOT HEARD ROUND THE WORLD

Turns out that in addition to insulin, we have other peptide hormones.

Among them are GLP-1, PYY, CCK, and GIP.

Together these activate insulin and the hormones that trigger appetite and satiety.

I call them the Switch.

Our Switch hormones by design turn off really fast, like a motion detector.

In the modern world, this presents a problem.

Scientists zeroing in on this modern problem found a prehistoric solution.

Gila monsters' venom contains a hormone like our Switch hormones.

But their Switch-like hormone lasts much longer, like a force field.

1   Meghan O'Hearn, Brianna N. Lauren, John B. Wong, David D. Kim, and Dariush Mozaffarian, "Trends and Disparities in Cardiometabolic Health Among U.S. Adults, 1999–2018," *Journal of the American College of Cardiology* 80, no. 2 (2022): 138–51, https://doi.org/10.1016/j.jacc.2022.04.046.

Enter the Shot—a Switch replacement that stays on longer.

The Shot helps people with diabetes optimize their blood sugar.

A side effect: weight loss.

Word spreads.

More than one in ten Americans have tried the Shot.

Many of them lose weight, optimize blood sugar, and are freed from food noise.

Most get the same plan: more protein, more strength training.

Inflammation and sleep apnea abate—as long as they stay on the Shot.

Many of those on the Shot experience physical side effects. Most power through.

For about 5 percent of those who try it, the Shot doesn't work.

Prohibitive side effects for others: lack of access and high cost.

Ultimately, the majority go off the Shot.

Most do it too early, without a plan.

They regain weight. Their blood sugar elevates.

YOUR BEST SHOT

You've got a goal: optimal weight health, today and tomorrow.

Your weight health depends on the status of your weight-health ecosystem—and specifically, that of your Switch.

To thrive today, it takes more than a motion detector. You need a force field.

With Switch Optimization, you build your force field—with or without the Shot.

CHAPTER 2

MY STORY

HOW I WORMED MY WAY TO OPTIMAL HEALTH

On the way to putting the RD after my name, I struggled. Made a lot of wrong turns. A few face-plants. I could have used a good GPS, but there was no GPS.

To write this book, I had to live it first.

In the decades that scientists were trying to devise medications that could address obesity, what is arguably the largest-scale health crisis of our time—some focusing on weight, others on blood sugar and cardiovascular and cognitive diseases—I was doing my own research. At first I had just one subject—me. What I learned from me, with the help of some great practitioners, took me to dietetics as a profession. Since then, I've learned a whole lot from my patients. (I kept studying myself, too, as I transitioned through life stages and health challenges.)

How the scientists, thanks to Gila monsters, got to the Shot we'll cover later.

How I got to *Your Best Shot* is a somewhat meandering story.

No reptiles—at least not the cold-blooded kind—were involved. There were some hoofed creatures. And a parasite that, looking back, I suspect was imaginary.

Growing up and in my twenties, I had a severely suboptimal Switch. Not that I knew it. Nor would you know what that means . . . yet. I thought my issue was being fat and everything society told me that entailed—overeating; eating the wrong foods; eating at the wrong times; eating, period. Eventually—and it took years—I saw that was wrong. I learned that what I was living and struggling through was a digestive-health story. And from there, it evolved into a weight-health story—where my weight type and location were key performance indicators for my health—and fundamentally a weight-health-ecosystem optimization story. But this was the '80s, and I was growing up in middle-America Ann Arbor, Michigan, then Columbus, Ohio, in a middle-class family, as the middle child: There was no ecosystem to optimize. There were dietary guidelines, scales, and antibiotics. By the '90s, there would be a lot of diet programs, too.

When I was growing up, my two brothers, parents, and grandparents didn't try to tell me that since I was "the girl" I had to act differently. With my brothers as my first role models and best friends, I played and competed like a boy; and when circumstances called for it, I fought like a boy. Consistent to a fault, I also ate like a boy.

Sure I had Barbie, Ken, and the Barbie car. But I also had G.I. Joe and played Dungeons & Dragons and Risk, seeking to assert total world domination. People called me "cute," and not a few pointed out that I was a Shirley Temple look-alike. For a run of my birthdays, the cake was a castle, I was the queen, and all the boys were vassals I knighted with cardboard swords and assigned quests and such. When one of the TV networks launched a show titled *Who's the Boss?*, I had the answer: Ashley is. I enjoyed being me. I thought I was funny, and people bought in. Or went along.

I was a standard-issue 1970s kid. I did establish a cycle of perennial strep and ear infections, but the antibiotics—always at hand—knocked them right out. I'd down the pills and launch myself back into the fray. About the time I was ten, life made a seismic shift. I went from being okay with being me to trying to be Angela Bower. The titular character of *Who's the Boss?* had it all: a cool job, a cool guy who stayed home while she went to the cool job—everything I dreamed about. And she didn't have the thing I suddenly discovered I did: a belly.

Shirley Temple had one. But this was the '80s, and bellies were out. Since I'd grown up with brothers, I'd been teased since I could walk. But the teasing changed. Meanwhile, adults responded to me differently—some were doctors—and the message was that the way I looked was no longer okay.

In high school I tried every diet out there. There were a lot of diets. Still, my belly didn't go away. I figured I could at least have Angela's blond hair instead of my natural brown ringlets that screamed *fat little Shirley Temple!* Sun In turned my hair orange; I fried it to a crisp with straighteners and then was instructed by my mom to cut it all off. Though the vivid term *hot mess* was a few years in the future, I was one. Focused 100 percent on my weight problem and the only known solution, dieting, I was on the wrong track. Since there were a lot of other people—millions—on it with me, I didn't catch on.

I arrived at college eager for discovery. My discoveries freshman year included pedicures and eating disorders. Always a diligent student, I studied people around me and even tried to follow a few of their programs. Laxatives, bingeing both exercise and food, then not eating, seemed to work for them but didn't for me. I was a failure at anorexia and bulimia. I *was* successful at binge drinking. Liquid courage helped me socialize and pretend not to care about my not-okay-ness. I veered further from better health.

Fast-forward to graduation. Diploma in hand, I made a stop back home in Ohio, where I got case 10,001 of strep throat and an ear infection. Same old. But not the same old treatment. Now a doctor labeled this as a chronic health issue. Solution: Remove my tonsils. The procedure went fine, but a postsurgical infection kicked in. Over the course of a month I ate, for all practical purposes, nothing. Even drank next to nothing. I saw only the upside: At long last, I got rid of that belly.

Being thin for the first time in my life was a literal out-of-body experience.

I out-of-bodied my twenty-one-year-old self to New York City and my dream job-life working at a big ad agency. Basically I *was* Angela Bower.

At work and play, I got a lot of attention. I was used to being noticed for my ideas, my quips, for knowing stuff, but the new flat-belly me was getting attention for another reason, the reason we all always deep down knew to be the one that really counts. There were days and evenings I felt like the fifth lead on my new fave show, *Sex and the City*. It was a novel feeling, but it wasn't a comfortable one. So, I leaned into it—into not eating and into my continuing success as a drinker. That regimen worked to get me past my qualms. But soon my belly returned. *But I wasn't eating! And, okay, I drank alcohol, but not nearly enough that the calories offset all the not eating I was doing. WTF?*

Anxiety set in, followed by panic attacks. And that took me back to my old standby comfort foods, holding my breath to get into my pants, and that self-view I believed I'd put in the rearview mirror: fat and failure. I went on a new round of fad diets, these more up to the minute, even New Age: Yoga. Macrobiotics. Doctors recommended what they had. Per them, there was nothing "wrong with me," but try birth control for those heavy periods. I heard "It's all in your head" and got myself a therapist. I ran the New York City Marathon (five hours and change, thanks for asking). Much to my dismay, after months of training and the hard-won 26.2 miles, I still had the belly. I quit alcohol and ate only after sundown (just like the pizza-loving client you'll meet in the next chapter, I'd inferred from all the diet noise that day-eating calories could be problematic), not counting the soy milk caffè lattes I slammed on an empty stomach for midday energy.

Even those efforts, individually and collaboratively, did little to improve how I felt . . . and looked. I can't quite say I tried everything. But I tried a lot. And *none* of it worked.

Then one afternoon, hanging like a bat upside down at yoga, I got to chatting with another inverted person: red-haired, super-fit, and super-confident. I must have groused. A healer, she offered to help me out.

In her clinic-apartment, peering at my blood under a microscope that I swear was branded Fisher-Price, she asked, "Did all this start when you were about ten or eleven?" We've all had that moment. That weird experience with a tarot card or palm reader who tells us something that just rings true.

She went on. "And has it worsened every year since?"

She could see *that* in my blood?

Scientist? Psychic? I didn't care.

She was taking time and connecting the dots in my story. She saw *me*.

She diagnosed the problem: A worm. A worm that had lived in me for years, and grown, and given me this belly. The worm was sapping my energy, ruining my sleep, and driving my insatiable appetite. I felt like the guy in *Alien* who's the first to go. I had to get this worm out before it came out on its own.

The healer was on the case. "Do you want to kill the worm?"

I sensed that this was like the question from the flight attendant to the exit-row six—it had to get a verbal, affirmative answer. "Yes, I want to kill the worm."

The way to kill the worm was to live for a week on nothing but goat's milk. Raw. Forty ounces each day, minimum. I did the math: Five cups did not sound excessive. On the other hand, it was more goat's milk than I had consumed in my lifetime. And how many goat milkings did it represent?

My pointless speculations were cut off by the healer. Goat's milk was not only the perfect food—it delivers carbs, protein, and fats—but also it was alkaline.

I read between the lines: Alkalinity is to worms what garlic is to vampires.

The milk would keep me nourished while exorcising the worm from my system. The healer's parting instructions: "Watch your poop. When it turns concrete white, you're passing the dead worm." I beelined for the one natural food grocery I was aware of in all of Manhattan and bought out their supply. I had seldom been so excited. Or so hopeful.

I'm telling you this part because it's nuts—nuts in a way that a lot of us know firsthand and have gone for hook, line, and sinker. After an unending saga of frustration and letdowns and false hope, we're ready to try anything. We've hit our personal worst, our rock bottom. Not that we even know it.

The worm-kill cure worked. Just as promised, I was full of energy and I slept well. My therapist noted that I was the healthiest she'd seen me. Testifying to all this was pale-gray poop. And a flat belly. Tempting as it was to subsist for the rest of my life on goat's milk, I followed the protocol. Day 8, I went back to my regular habits. Ding-dong, the worm was dead.

And in a week the belly came back. Maybe the worm was immortal.

My bathroom scale corroborated what I knew: I'd gained back the weight I'd lost and a pound for good measure. A pound to punish me—for my hubris and arrogance. Or maybe it was the worm-god's revenge. Where to go for solace? The Liquor Store— uppercase because it's not a retail outlet but a watering hole in Tribeca. Big with day drinkers. I bellied up to the bar. We took turns telling stories. My goat's-milk-cleanse story got a good laugh—at me *and* with me. Before I got to the end, the listeners raised their glasses and chanted, "Yes, I want to kill the worm!"

A guy seated on his own down the bar looked up from his burger—at me—and asked, "You take antibiotics as a kid?" Huh? Was this guy from a different century? "I grew up in the '70s and '80s. We freebased them." He just nodded. That sage nod of those who know much and speak little. I added a nervous bit on the serial throat and ear infections. "Yeah," he concluded, shifting on the barstool, to look my way. "There's no worm. But your body also doesn't have any good bacteria."

The spokesman for good bacteria turned out to be a doctor. A different type of gastroenterologist than I had ever met. When I consulted with him a few days later at his practice (in an office,

not an apartment), we discussed the poor health I'd been experiencing for, well, my whole life: gas, bloating, constipation or
diarrhea (what I will later discuss as the Goldilocks of motility
issues), ear and throat infections, and, more recent, severe panic
attacks. Weight was not on the list.

My belly, I began to grasp, was a *signal* of all these issues. My
belly was my body telling me—okay, screaming at and begging
me—to pay attention to it from the inside, not the outside.

More impactful than any specific recommendation he gave me
was his approach. He engaged me with curiosity, asking questions
about my history, my health, my choices, and me as a whole person. This complemented and informed his diagnosis, which was
that my gut was not getting what it needed to run better. It was
having a really hard time doing its job. Its *jobs*, as I was learning.

This would be a defining piece of my story: The hot mess was
my gut, not me.

"We've overprescribed antibiotics for a long time, and they
wipe your system's good bacteria—probiotics—which upsets your
digestive system's overall function," the GI doctor said.

Over three months, we introduced probiotics and I resumed
my macrobiotic diet, which collaboratively improved my gut function. For the first time, I felt and looked better, not because of
what I was avoiding—all nutrition except goat's milk—but because I was giving my body what it needed. The connection of digestion to weight health was cemented for me. Why had no other
doctor ever asked me about my digestion before? Those questions
and a bookstore search (this was pre-Google) ignited my mission
to help others get to these answers before a rock bottom like a
goat's milk cleanse.

The doctor didn't *do* this so much as he oriented me toward learning how to do it. Soon I was reading and sitting in on talks about food, supplements, breathing, and more. Especially about the gut bacteria. In those days, civilians like me hadn't even heard the term *probiotics*. It would be years later that I would discover the not-so-conditionally essential role of glutamine (a.k.a. "the glue").

I was learning how to heal. Learning that what was in order wasn't a one-time, take-it-into-the-shop fix but an ongoing practice—continuing assessment and optimization. Just as important in the long run, I was starting my practice—with myself. A lot of people were out there, ahead, waiting for me, and I was getting up to speed. I would leap tall weight-health issues in a single bound, help other people fix their digestion, and—maybe more empowering yet, to me as much as to them—enable them to skip, once and for all, a goat's milk cleanse.

Healing my digestion magically cured my anxiety and panic attacks. But not magically at all. I came to see and feel the reciprocity between my physical gut and gut feelings—and other feelings, too. These changes upped my energy and, yes, reduced my belly. As a bonus, they finally eliminated the proverbial worm of my I'm-not-okay need to be Angela Bower. It would still be some time before I'd be 95 percent with my size and shape, but I was really good with being Ashley, which was a new out-of-body-but-in-the-body experience. I had not felt centered in myself since my age was a single digit.

After doing the requisite digestive repair work, I dived into personalizing my plan. My own experiments and lab tests told me that my body is allergic to barley (no wonder my belly revolted when I drank beer) and that it doesn't do well with milk from a cow *or* a soybean. To boot, it metabolizes caffeine very slowly. So all those soy lattes with two shots of espresso I knocked back to get through the day were making it harder to get through the day. On the upside, hemp seeds are for me a true superfood.

And finally, my motility—how my digestive tract moves—was slow, and even slower at certain times of the month. This discovery proved critical in building my understanding of not only what foods to choose but also what nutrients and activities I needed to help my digestive tract pick up its pace. This plan built on personal intel, repeat assessment, and updating choices—optimization— enabled me to enjoy an excellent health span during my thirties and forties.

I mapped my plan as I went via the following steps, and in doing so, I laid the groundwork for a system that enables you to optimize your weight health, too:

- Assess and optimize digestion.
- Identify *my* better nutrition choices.
- Develop a lifestyle plan to complement my nutrition.

What I've built since that fateful meeting at The Liquor Store is a blueprint for the most efficient and effective route to optimal health span: *your best shot.*

KEY TAKEAWAYS

- Never do a goat's milk cleanse. It won't kill any type of worm—real or imagined.
- Optimizing digestion is foundational for all health goals, especially weight.
- There are a ton of tools—supplements, protocols, activities, saunas, cold plunges, and yes, the Shot—purporting to

deliver weight loss. For weight health, many of them are actually better. The better ones for you? That answer requires assessment.

THE TAKEAWAY

Our weight-health story is ours—written, operated, and lived by us.

What your weight-health ecosystem needs and which choices give you your best shot—that's what's next.

CHAPTER 3

PATIENT STORIES

WHAT THE SWITCH NEEDS

"Good luck," the doctor wrapped up, adding, "To be honest, I'm not expecting you to succeed. We have to show him we tried. He's one of our major donors." Today, this candidate for bariatric surgery would be on the Shot. Even so, I'd likely still get called in. His case was complicated, and his digestive complaints revealed a challenge that often affects how the Shot works. At his house, he walked me through his current regime—his "last-ditch effort." He wasn't losing, but at least he wasn't gaining. His diabetes was getting worse, though. "I don't get it," he explained to me. "You may think I'm crazy. I have nothing but diet soda all day, till around 7 p.m. Then I have my pizza. I get a plain cheese, and I eat it until about an hour or two before bed. I chose it because I was told not to go too low-calorie since I gained a lot of weight on a really low-calorie diet."

The fact was his food choices were not crazy. The calories in his nightly large pizza matched the number in his daily nutrition recommendations, and the breakdown of protein, carbs, and fats was all right. He was ordering from a place that made terrific pizza from high-quality ingredients. I asked about his digestion. He reported being bloated and gassy all the time. From his doctor, I knew that he was on medications for acid reflux as well as high blood sugar and high blood pressure. (This was not only pre-Shot but also pre-statin. Today, I am sure he would have been on a statin, too.) He battled depression, so he took medication for that as well.

We reviewed all his weight-loss efforts. He truly had tried everything. The last thing he needed was another prefab, off-the-rack, destined-to-fail plan.

He needed a plan built for him.

We worked one up together:

Cut out the diet soda.

Coffee or tea not more than twice a day.

8–10 ounces of water every three hours (once each day, add electrolytes).

Back then, we didn't have data on the gut microbiome, or even the term, but I suspected that medications had interfered with his. The diet soda was adding artificial sweetener to the mix, messing up the microbes and confusing his sweet taste buds. All this spoke to the need for digestive support via nutrients. So, we worked in glutamine, magnesium, and bifidobacteria (a specific strain of probiotic, relatively new at the time).

The centerpiece of his plan—the hub of personalization that made it *his* plan:

Stick with the pizza.

He was stunned. His doctor was stunned-er, and I got calls and pages to tell me so, loud and clear. My thinking was this: The issue to tackle first wasn't *what* but *when*. Timing. This, along with optimizing digestion, was his better next step. The body, his and yours, is designed like a race car, not a passenger car. It will run better—blood sugar, energy, digestion, releasing fat from storage—if it gets what it needs *when* it needs it. He'd been backloading calories into the last four hours of his day. We redistributed the same calories over the run of the day. Swap out one big meal; swap in five pit stops—my term to refer to anytime we give our body nutrition (meal or snack, food or liquid)—for refueling: two to three slices of pizza, every three hours, starting an hour after waking and going on through to 10 p.m. (three hours before his standard turn-in time).

At the four-week mark, he'd lost twenty-eight pounds. Finger pricks said his doctor could reduce his blood sugar medication.

At six weeks, he was down another twenty-three pounds.

Total weight loss: the fifty he'd needed to lose plus one more.

The doctors cleared him for the life-saving surgery.

Mr. Pizza Pop—we all loved the nickname he chose for himself—was weight healthier, on his terms.

KEY TAKEAWAYS

- No one is doomed to weight-health failure.
- The foundation of weight health: better digestion (including hydration).

- Fat loss and optimal blood sugar require better nutrition.
- Better nutrition better be personalized to your body, today.
- Every personalized plan optimizes better nutrition's four pillars:
  ⬧ Balance, Quantity, Timing, Quality
- A personalized plan addresses the pillars differently.
- Mr. Pizza Pop needed to improve timing and quantity:
  ⬧ The old rule—X calories a day—doesn't work.
  ⬧ *When* we eat matters, not just how much.

WHAT THE SWITCH WANTS

Tammy and I met and decided no, she was not going to be a Shot failure. "These are the same fifteen pounds I've gained and lost on other diets," she told me. "On tirzepatide I just lost them again." She ran me through all the things she'd been advised to do and had been doing: exercise, eating whole foods (mostly plants), not overeating (most of the time), and taking rigorously curated supplements. "I'm stuck," she said. Far from being noncompliant, she was doing, even exceeding, all the Shot recommendations—which I realized was the problem. An overachiever, Tammy was checking every possible box. Faithfully following the tirzepatide plan, she was going with recommendations that didn't fit for her, and she was missing a few that were essential. What she needed was the Tammy plan, which we went to work on building. Since she was trying everything, we had plenty to experiment with. Our starting point was a comprehensive assessment, going beyond "Are you exercising?" to types of exercise, how often, how long; beyond "What's your protein intake?" to "Do your meals and snacks deliver at least 20 grams of protein, and how much carbohydrate?"

Plus, the foundational ones, "How is your digestion?" and "How's your hydration?" Those assessment points gave us enough to figure out how to *un*stall her. The first goal was to move beyond that fifteen-pound plateau by adding fat loss (no muscle included).

Here's the Tammy plan:

My scan of her choices in the context of the four pillars of better nutrition identified *balance* as the first step. She was getting in a good amount of protein but split between lunch and dinner. Breakfast and snacks had insufficient protein intake. Dinner was too high in carbohydrates. (Her plant-based protein sources were also carbs—good-quality, high-fiber ones, but they netted out as too much carbohydrate at one time, especially her last meal of the day.) All this, along with her morning 16-ounce kombucha, and I could easily explain her elevated triglycerides, too.

With better balance as the goal, she set out to experiment with new combinations of her choices. For breakfast, a protein drink or nondairy yogurt with hemp seeds and berries, then a protein drink after her workout. Lunch was a variation of a ½ cup of beans, 1 cup of bone broth, a combination of non-starchy vegetables either as a salad or sautéed, and avocado and chicken or tofu. Her midafternoon snack was a bar that she loved, which included 17 grams of protein, 15 grams of carbs, and 10 grams of fat from mostly plants with some whey protein. Dinner was now fish or a veggie burger—no bun—with vegetables in olive oil and a variety of sautéed non-starchy vegetables. She loved her nondairy cheese, so that was dessert.

All these experiments, designed to optimize nutrient balance, could do only so much without optimal digestion, which includes hydration—not just drinking water but absorbing water and nutrients into the cells for metabolic benefits.

Tammy self-reported that she peed All. The. Time. Her hydration experiment—a key part of GLP-1 optimization assessment, which you'll see on page 177—confirmed we needed to improve her absorption of water and the nutrients it carries into the cells. So, we added electrolytes to her water following exercise, and we reduced dehydrators. To optimize digestion, we incorporated digestive enzymes to improve the breakdown and use of protein, especially from her preferred plant sources. Liquid chlorophyll in water with lemon replaced her morning kombucha. Kombucha's bubbles added to her bubbliness (the gas and bloating did not add up to a positive morning mood), and its sugar contributed to her carb overload. But the water's chlorophyll and lemon supported her microbiome to make it more hospitable for the good probiotics. Plus, it gave her a way to tolerate water, since her low intake was tied to disliking the taste of plain water. Tammy had a history of elevated cholesterol (LDL, ApoB) and inflammatory markers (hsCRP). She also had genetic variants that signaled the need to make it easier for her body to turn off inflammation. I recommended resolvins (pro-resolving mediators) and omega-3s as supplements, too. Weight health can't occur when the body is distracted waging war. The inflammatory response is a key tool the body uses temporarily to alert and address issues; it needs to be turned off—resolved—when situations are dealt with. The final adjustment was *how* we tracked her results. Rather than measure body composition at a facility once a month, she monitored her body composition weekly on an at-home scale, and we rechecked her cholesterol and inflammation labs every three months. Two weeks into Tammy's new plan, the scale showed fat loss without any muscle loss. A month in, her inflammatory markers were in the optimal range.

In the months that followed, Tammy added caloric reduction in solidarity with a work colleague also on a weight-health journey. Reducing her weekly intake by a total of 3,500 calories, Tammy was doing everything she could, and it worked. She continued to lose fat while gaining muscle. By the four-month mark, she'd lost an additional twenty-five pounds—on top of the plateau fifteen—of visceral and subcutaneous fat. She'd added lean body mass. All of her labs were at optimal levels. And her quiet, efficient digestion told her that her body liked her new choices. That got a big thumbs-up from her family! Tammy set a new goal: For the first time in her adult life, she was aiming to be under 25 percent body fat. Game on!

Tammy had never been predestined to be a tirzepatide failure. Did she *need* the Shot to achieve what she did? Tirzepatide was part of what she was doing, part of her plan, so that's a hypothetical we don't need to answer.

What does *your* body need so you can make the choices that will help it run better? That's what we will answer together.

KEY TAKEAWAYS

- Better body composition, not weight loss, is our goal.
- The Shot is a tool that works better with a personalized plan.
- The Shot can stop working; and for some people, it doesn't work at all.
- Better nutrition optimizes the weight-health ecosystem, Shot or not.
- Tammy's plan optimized quantity and balance.
- Optimizing hydration and digestion provided foundational ecosystem support.

HIS BEST SHOT

Don came to me to lose twenty-seven pounds. He was exact: "I need to get back under two hundred." His doctor mentioned the Shot, but he and his wife were concerned. What did I think? Over our session, starting with a comprehensive assessment covering digestion, hydration, and health history, we discussed the nutrition and lifestyle changes that I thought would help him reach his goals. The problem: At some point he had tried almost all those changes with varying degrees of short-term success. In efforts to achieve goal levels for cholesterol, blood sugar, and blood pressure, Don was already on a lot of medications. His diagnosis of metabolic syndrome, along with chronic inflammation, played a key role in my next suggestion: "A weight-health-hormone replacement as a tool right now looks to me like your best shot. We can use it to help you shift choices and optimize metabolic function. Without it, we *can* do the same, but you've been there, tried that. I say we do this as one part of a personalized weight-health plan. We can keep your dose low—using the Shot as a tool, not the solution." After we considered more of their concerns, they were both on board.

His wife was fully ready to help him implement the nutrition recommendations. To support him, she agreed to three months of no alcohol, along with Don. Since he reported that making time for strength training was hard, I connected him to a trainer who could be flexible about scheduling and would be available to him virtually for sessions.

Six months in, Don was at a weight composition he's thrilled with: a tad over his original two hundred but with more lean muscle mass and the return of collegiate-level abs. He's also off all medications other than the tirzepatide. As a team, Don, his wife, his doctors, and I agree: It's ideal for him to stay on the Shot and maintain these levels for a year. Will he go off the Shot? Maybe. He may also use a maintenance dose longer term. We will continue to evaluate as his life happens.

KEY TAKEAWAYS

- People benefiting from the Shot *have* tried diet and exercise. The Shot is a tool that can help them where past efforts haven't.
- Weight composition (fat, muscle, and bone mass) is a key performance indicator but so are several metabolic health labs.
- Your best shot is a personalized plan that is doable for you, today.

SWITCH, NOT SHOT

As Stacy ran down her laundry list of symptoms—epic fatigue, debilitating back pain, and about fifteen pounds of post-babies belly weight gain that she couldn't lose no matter what diet she tried—dots started connecting. As I listened for clues, I heard "acne" and a medication for it. Those answers pointed to a dual issue. Acne is a signal that digestion is suboptimal, and medications taken to treat acne disrupt digestion, the location for deployment of our weight-health hormones.

Among her goals was "to be able to keep up with my kids." Another: "I don't want to use Ozempic. My insurance doesn't cover it. We can't afford it, but even so, I just don't want to use it. Is there anything I can do that will work without it?"

We started with this set of baseline actions:

- Optimize digestion and hydration with tune-ups via adjustments to her supplements: add glutamine; change her current magnesium to a form that's better absorbed; change her probiotic to add GLP-1-supporting *Bifidobacteria*, *Clostridium*, and *Akkermansia*; add an electrolyte.
- Optimize nutrition with a focus on timing and balance. She was often noshing until 9 p.m. or 10 p.m., depending on her husband's schedule or her post-kid-bedtime treats while she finished up her daily to-do list. We closed her caloric window at 8 p.m. (We closed her work and screentime window, too.) To achieve better balance, we adjusted her meals to deliver 15–30 grams of protein, at least 5 grams of fiber (including a daily dose of 2 teaspoons of ground flaxseeds), and no more than 30 grams of carbohydrates per meal or snack. We also added a daily serving of GLA (the "glamour" fatty acid) from wild salmon, hemp seeds, or a supplement.
- Her vitamin D was low, so we increased her daily amount with a supplement containing K2 to increase vitamin D absorption and promote bone health.
- The over-the-counter medications she took most days for back pain are weight-health challengers, so we replaced those with omega-3s to prevent inflammation and resolvins (pro-resolving mediators) to turn it off.
- We changed what she tracked—not just a number on the scale but her waist circumference and waist-to-hip ratio. Later she added a scale that measured body composition (pounds of fat, skeletal muscle mass).

This was Stacy's plan. I was confident it would optimize her weight-health ecosystem—contingent on her ability to implement it. However, I had a hunch I was giving her a plan without a tool to help her tackle her cravings and food noise. Enter the natural GLP-1 activator, a New Zealand hops extract (Amarasate®). No, you can't get it in beer.

Stacy noticed differences almost immediately. Within a month, she was down five pounds, but her husband told her it looked like fifteen. She went away for a girls' weekend, ate and drank what she felt like, and continued with fat loss—"and zero guilt."

After one hundred days, we celebrated. Stacy's weight-composition changes included three inches lost in her waist, twelve pounds of fat loss, and two pounds of muscle gain. She attributes this primarily to being more in control of stress and emotional eating. Her digestion had improved in a big way and resolved issues of gas and bloating. She noted brighter, more vibrant-looking skin and stronger nails, and near total freedom from monthly cold sores. Her periods were regular, too.

All of these outcomes were exciting and wonderful, but the best was no more back pain, which meant she could trek with her kids—including carrying the little one on her back—increase her daily exercise, and attend to all of her to-dos without pain medication. In Stacy's estimate, her biggest win was "I am a better mom."

KEY TAKEAWAYS

- Medications can work for certain issues *and* be weight-health challengers.
- Amarasate®, a GLP-1 activator, can be a game changer in the battle against cravings and food noise.
- Personalization is how you optimize weight health . . . Shot or not.

WHEN TO STOP THE SHOT . . .
IS NOT THE BETTER QUESTION

In our initial conversation, Jan cut to the chase: "How do I go off this successfully?" I came back with a thumbnail of my three-stage process: (1) comprehensive assessment, (2) optimizations as indicated, (3) tapering off the medication with tracking. Jan's takeaway was that rather than cold turkey as she'd pictured, we'd phase it out.

Jan's phase 1: Her assessment provided data that would impact her long-term results. She had lost weight—pounds on the scale—but prior and recent DEXA scans—a body composition test—revealed a troubling situation. She had lost muscle and bone mass, too. This was currently influencing her metabolism, and it was a red flag for challenges off the Shot. Her digestion and hydration were considerably less than optimal. Her plan needed to address all of these before we began stepping down the medication.

With regard to nutrition, we needed to optimize her protein. She was getting in 100 grams of protein daily, but several of her pit stops (meals and snacks) didn't deliver a minimum ecosystem-supporting quantity. Her meals that were high in protein were too high, which contributed to digestive complaints and provided a risk factor for weight gain post-Shot. For muscle building and retention, we targeted her amino acid intake, adding creatine with taurine. I suggested she start strength training at least twice weekly. She went right to it, joining a class her friends attended.

For digestive optimization, we swapped her probiotic gummy for one with *Bifidobacteria*, *Akkermansia*, and *Clostridium*. This noncaloric, unsweetened combination of strains supported her gut health, specifically targeting GLP-1 hormone activity. That helped lay the foundation for resuming her body's production of this key hormone, which the Shot suppresses. We added glutamine to promote the integrity of the mucosal lining of the intestines—where those digestive hormones reside—and the formation of muscle.

Finally, we looked at hydration. Since she'd started the Shot, her water intake had dwindled, as I've commonly observed. In her case, she only drank water while exercising or just after, when she "chugged about 24 ounces." We shifted to a pit-stop approach: regular intakes of 8 to 12 ounces every three-ish hours. (She set an alarm as a training-wheels tool for this one.) To her post-workout water we added magnesium, an electrolyte, to help hydrate, turn off stress from the workout, and support bowel regularity.

We started weekly weight-composition checks—not daily weigh-ins—on a specialized scale to give us personalized, actionable data.

After a month, the undifferentiated number on the scale—weight—was unchanged. But the meaningful numbers told a different story. Jan had lost two more pounds of body fat and had put on two pounds of muscle. It was hard for her to see around the zero-sum interpretation. Since forever, she'd been brainwashed with one message: *Lose weight*. Losing that mindset can be the harder journey.

Once Jan saw that she had made progress, her question was: Could she start coming off the Shot?

For sure—in a careful step-down. Since the medication suppresses weight-health-hormone production, we had to ensure that we stepped it back up. She stuck with the same dose but took it less often. We added the GLP-1 activator Amarasate® to provide additional support on the days between Shot doses. This powerful, natural go-between, taken an hour before eating, delivers a meaningful increase in natural weight-health-hormone production, resulting in a reduction in cravings and calories consumed. The effect lasts about four hours.

At nine weeks out, with things holding steady, we reduced the dosage again.

At the three-month mark, Jan still showed good progress. Her numbers stayed stable, and her body composition improved. Going off the Shot altogether looked like a good next move.

At that point, Jan hit a setback—a major upset rocked her life. In reaction, she put on eight pounds, which I learned about from her therapist because Jan dropped contact with me; she'd written herself off as a weight-health failure. I knew it could feel like backsliding for her, but I suggested through her therapist that her best shot was to safeguard the gains she'd made by going back

Why use the Shot if there's nature's Ozempic? I like your thought process. So do supplement marketers. Fact is, despite what you see and hear, there is *no* nature's Ozempic. There are, as you will learn, many supplemental ingredients that support GLP-1 optimization, and there is at least one—Amarasate®—that activates weight-health-hormone production beyond typical baseline levels. However, weight-health hormones like GLP-1 produced by your body are deactivated by enzymes deployed as the hormones are secreted, to break them down within minutes. The unique configuration of the Shot is that its version of the hormones does not get deactivated for nearly a week. (The time frame depends on dosage, ingredients, and an individual's body.)

to the regular weekly dose for a while. She was eager not to spiral down, and her doctors and therapist both supported the move. Rather than go to her prior highest dose, though, she kept with the activator and stayed at closer to her starting dose.

A few months later, the upheaval in her life dealt with, we went back to the phase-out.

Her assessment revealed that everything was optimal, so we were able to space out the current dose to every fifteen days, and then just once a month. At nine months, she was off the Shot, down another fifteen pounds of fat, and up two more pounds of lean body mass. Her bone loss staved off, too.

She keeps a dose of the medication in her refrigerator as a reminder that it's a tool available if she decides to use it. If she makes that choice, she has a plan for how to use the Shot, whether at full dose or lower doses and frequencies; how to phase off it; and how to optimize her weight health.

For now, by the state of what we know, she's got all weight-health bases covered.

KEY TAKEAWAYS

- On the Shot, off the Shot, or transitioning from one to the other, the structure that supports *everything*: GLP-1 optimization.
- Your story is *yours*, not a set prescription plan. When it looks hopeless, there is always a pivot, an optimization.

PART II

WHAT'S YOUR BEST SHOT?

CHAPTER 4

YOUR BODY *IS* AN ECOSYSTEM

A weight-health ecosystem. This is *not* a metaphor. It's a thing. And for the sake of conversation, we'll shift to calling it "the ecosystem" or "your ecosystem." There are various ways of describing an ecosystem. I go basic. An ecosystem is a community of organisms (living ones) of all kinds—plants, animals, microbes, fungi, blue-green algae (and possibly others)—*with* all the living things' interactions and interdependence, both among themselves and with their environment.

Your weight-health ecosystem—a.k.a. your ecosystem—is your body. All its systems, parts, pieces, and critters. Some worth highlighting for their direct relationship to your Switch and its optimization work—and the six we'll focus on—are:

- Digestive (subsystems: motility, breakdown, absorption, hydration)
- Nervous (star player: the vagus nerve)
- Cardiovascular

- Endocrine
- Musculoskeletal
- Detoxification

Those are your weight-health ecosystem's core component systems. All ecosystems are defined by interactions. The interactions that make up your weight-health ecosystem initiate chains of events to complete missions, day and night. Here they are described, along with their Switch Optimization roles.

DIGESTIVE SYSTEM

Digestion is the set of processes and reactions whereby the body receives, evaluates, processes, disseminates, and eliminates what it receives via intake. This system is composed of the mouth, esophagus, stomach, gallbladder, liver, pancreas, small and large intestines, and anus. Notably, your Switch hormones are secreted by cells in the lining of the intestinal tract in response to what is introduced or by the nervous system as the brain senses nutrient opportunities to manage. This system is ground zero for Switch function. Any slowdown, acceleration, or damage here impacts what the cells get, what gets stored as fat instead of put to use, and what is deemed problematic and not to our benefit. Here are its subparts, including hydration, which is often not thought of as part of digestion, but as we'll see at various points on this journey, it belongs here:

> **Motility:** The digestive system's need to move food and nutrients along from one organ to the next makes motility (movement) a primary task. This movement is controlled by

four main factors—nutrients, gravity, contraction, and the nervous system's input. Your Switch hormones impact and are impacted by the motility of your digestive system, thus we assess and repair as indicated for Switch Optimization.

Breakdown: You primarily consume food as liquids or solids, not nutrients. The organs of the digestive system combine to break down food into usable nutrients. This requires optimal amounts of fluids, enzymes, and microbes at the right locations. Your Switch hormones direct actions that impact the breakdown of nutrients, so we assess and optimize breakdown regularly.

Absorption: Having nutrients is no benefit unless they get to where they are needed. Your digestive system lining decides what gets absorbed and what gets targeted for elimination. Since your Switch hormones are produced by cells in the lining of the intestinal tract, assessment and optimization of the lining will be critical for Switch function and for the upgrade of your weight-health ecosystem.

Hydration: Hydration could be called the other nutrient-absorption process. Whether you are a hose or a sponge, as you will learn in your assessment on page 102, the efficiency of your hydration dictates how the nutrients that your ecosystem receives and processes will be absorbed into the cells. Your Switch hormones affect your hydration status as their responses to signals received in the colon trigger hydration actions—the absorption of water and electrolytes. Beyond how much water your body has, where that water is located is a key part of how your weight-health ecosystem operates.

NERVOUS SYSTEM

The nervous system is comparable to the digestive system in both the number and caliber of operations it's involved in or responsible for. These two systems interact continuously in a symbiotic relationship, particularly via the *vagus nerve*, making their relationship critical to your weight-health ecosystem.

CARDIOVASCULAR SYSTEM

The cardiovascular system includes body parts—heart and blood vessels—that are a target destination for and a conveyor of Switch hormones. Receptor sites exist on the heart muscle and blood vessels, and some Switch hormones travel via the bloodstream. For effective Switch communication and reception, the components of this system need to be fully operational—not too crowded, slowed down, or otherwise damaged.

ENDOCRINE SYSTEM

A network of glands and other organs, the endocrine system receives messaging from Switch hormones as they are delivered to their receptors. This interaction transfers responsibility for subsequent interactions to the hormones and fluids secreted by these organs. Adipose tissue (fat) is also a part of this system. As we would expect, fat cells contain receptors for Switch hormones. How much fat we make, what type, and the ability to burn it are all strongly influenced by the interactions of the Switch hormones.

With suboptimal Switch function, many or all of these

interactions are unsuccessful. The result is a cascade of operations challenges and failures.

MUSCULOSKELETAL SYSTEM

With the exception of bone density, the musculoskeletal system and the importance of optimal levels and quality of muscle and bone have too long been overlooked in weight composition and health span (and thus weight health).

A breakthrough opportunity of modern times is our ability to live longer, which becomes much more advantageous when we strive to be *healthy* longer—and gain a greater health span. To achieve that, we need optimal muscle and bone health. And yet weight loss and modern medicine have driven us to the opposite—defining better health as a lower weight regardless of body composition.

The specialty of obesity medicine and the measurement of our obesity crisis uses total weight or total weight for height (BMI—body mass index) as its primary marker. The lower your weight, the less risk for disease—right? Wrong! As my colleagues Dr. Gabrielle Lyon and JJ Virgin, CNS, CHFS, CPT, brilliantly teach, weighing more as we age due to maintaining and even acquiring lean body mass is essential for a body to perform optimally, especially for longer. Turns out, building muscle also helps to resolve—that is, turn off—inflammation. Switch hormones impact muscle and bone tissues getting what they need to develop and retain optimal function. Yes, fat mass matters, especially visceral fat mass. But if our objective is optimal weight health and optimal health span, we can't achieve that loss at the cost of a loss of skeletal muscle mass.

DETOXIFICATION SYSTEM

Detoxification is a multiphase system made up of the liver, kidneys, and skin to ensure the body processes and eliminates toxins. The body consumes and creates toxins constantly, even when operating optimally. The modern world has introduced myriad new toxins. To the body, many of these toxins look like resources it can use. Others it identifies as not for use and thus stores them in fat cells. The quantity and array of toxins can overwhelm the ecosystem, and specific toxins disrupt function. As we pursue Switch Optimization to restore your weight-health ecosystem, we need to ensure optimal support for detoxification efforts. One key reason is that when a goal is to shrink fat cells, which releases toxins stored in fat, we need effective elimination to avoid the toxins causing problems. Another reason is that brain cells, especially as we age, require optimal detoxification to maintain their integrity.

Each piece of your weight-health ecosystem deserves attention, and each will get it as you optimize your Switch function.

What ties them all together? Your ecosystem has a communications component. I've mentioned it a few times, but let's really dig into it. It's the Switch: the set of hormones that regulates your ecosystem.

The Switch directs interactions and chains of events to ensure that the body uses resources effectively and efficiently.

When we support optimal Switch function, the body gets more of what it needs to upgrade our weight-health ecosystem, resulting in better health today and tomorrow.

When the Switch is able to deploy and make use of what it receives in a timely and effective manner, the results are:

- Optimal subcutaneous fat
- Optimal visceral fat
- Optimal muscle mass
- Optimal bone health
- Optimal digestion
- Optimal hydration
- Optimal cardiovascular health
- Optimal detoxification

When it can't, fully or partially, the results are:

- Excess subcutaneous fat
- Excess visceral fat
- Insufficient muscle mass
- Bone loss
- Impaired bone growth
- Digestive complaints
- Cravings and appetite challenges
- Suboptimal energy
- Impaired cardiovascular function

Conventional medicine and health care recognize the ecosystem's components but only as stand-alones, not as a collective working in concert; it evaluates and addresses systems singularly, not as a whole.

Let's look at just one story that illustrates this. From our earliest schooling, we're taught we have a nervous system. Medicine treats nervous system dysfunction with medications that target the nervous system only and with psychotherapy. This focus—eyes on the trees, not the forest—results in suboptimal outcomes. Antidepressants are designed to regulate serotonin levels for better mood. However, we now know the majority of serotonin receptors are in the digestive tract! At the same time, those medications challenge Switch function. That's not to say that antidepressants don't have an important place in many personalized plans, but we need to look holistically at their impact on *your* ecosystem. If an antidepressant leads to bad weight health, it will be depressing.

WHAT HAPPENS IN VAGUS DOES *NOT* STAY IN VAGUS

The polyvagal theory, recently developed by Dr. Stephen Porges with clinical implementation support from Karen Onderko, gives us a better understanding of the scope of this awesome part of our electrical system. For me, it confirmed the identity of the Switch as an actual "switch" rather than just a set of hormones. My colleague Dr. Navaz Habib further helped me connect the dots of vagus nerve upgrade as an essential part of Switch Optimization.

The vagus nerve, which runs between the digestive tract and the brain, juggles many jobs, including the transmission of information in both directions. It is the primary communication channel linking the brain, the gut, and the rest of the weight-health ecosystem.

Today, the brain and the digestive tract receive and send more messages than ever. We eat almost around the clock and take in infinite stimuli via all our senses. This overwhelms a healthy digestive system and the vagus nerve. Increasingly the messages are difficult to decode or can't be decoded fast enough for the meaning to matter. On receipt, the brain and gut dutifully attempt to interpret and assign priorities to each other's messages. Think old-fashioned telephone switchboards trying to manage modern-day communications, both in volume and form.

Gridlock. Delays. Failed deliveries. Fatigue.

The digestive tract feels it.

The brain feels it.

And we feel it.

Adding to all this insult is injury. We've not only overwhelmed our vagus nerve's functions but also harmed the vagus nerve itself.

One of the main duties of a nerve? To feel.

The chaos of modern consumption and living—and all the associated stressors—makes the vagus nerve feel unsafe more often than not. Its functions get impaired and it gets damaged. Of course that's a problem, as those functions include delivering messages to trigger the Switch and delivering the Switch messages, once triggered, to receptors in the brain.

Because of this, your Switch Optimization includes an "upgrade [to] your vagus nerve," to echo the title of Dr. Habib's book. Because this is ecosystem work, literally everything we do will have an impact on digestion and the vagus nerve.

Sometimes what your vagus nerve needs to feel safe is a trusted guide. Throughout this book, you will be assessing, experimenting, and optimizing on repeat. Additionally, you will be encouraged to *un*learn well-established truths that turn out to be false. You'll want to offload them because they interfere with your weight health. Sometimes you (and your vagus nerve) may not feel particularly safe. In anticipation and for navigation purposes, I'll jump in with what I call GPS Alerts. They'll range from items for consideration to advisories that you may be heading in a not-better direction to outright blaring alarms and warnings.

The right guide makes all the difference for optimal outcomes.

GPS ALERT

FLY-FISH OR FLIGHT

I once lucked into the most brilliant fly-fishing guide. He made it fun, doable, and, to the extent possible, "choose your own adventure." Even for me, a person with a serious snake phobia. (Phobias, btw, take the downregulating effect of fear to next-level vagus nerve disruption.) This guide was, by coincidence, my brother.

A particular outing began with a potentially game-ending, vagus nerve–disrupting sighting of a snake (he told me later it was a copperhead) right there in the river we'd come to fish.

As my body shot to 110 percent fight-or-flight, my guide-brother calmly looked at me and, with 110 percent nonjudgment, said, "Okay, we can go. Let's just walk out right up here." He pointed to a sandy part of the riverbank. Then he said something that made me laugh.

To get to the river we had driven two hours. This was his one day off for a month. Yet his calm permission to activate the flight helped me feel safe. And it was the laughter that gave my vagus nerve permission to resume function.

The biggest surprise of his life, he says, was my reply: "I think I'm good. Let's fish." With my vagus nerve steadied, I was ready to go on.

YOUR GOAL: WHAT IS OPTIMAL?
AND WHY OPTIMIZATION?

We're all here—you, me, my patients you've already met, others you'll meet up ahead—because we have at least one shared goal: to be healthy and stay healthy for the duration, for the time we get to exist.

But how we regard this goal and talk about it plays a huge role in how effectively we achieve it.

With life-span and longevity, the goal is living longer. Simply living longer, while it has its perks, can come with burdens, too. Depending on how unhealthy we are, this is also burdensome for those we care about and for us.

The term *health span* better addresses our goal representing the time we're alive *and* enjoying good health.

The vehicle to help us achieve greater health span?

Enter Medicine 3.0, which ditches the reactive, fix-it model in favor of proactive and personalized health maintenance and health care.

Closer, yes, but not a bull's-eye.

What Medicine 3.0 doesn't deliver?

How.

Most specifically, how to resource *your* body for health span.

Resourcing is our action of making choices and delivering the resources that our body can consider and select for use.

Optimal, optimally resourced, optimization: Those three together are your money shot.

Optimal means the *best*, the *most* favorable outcome for the given situation.

Optimally resourced means your body has what it needs when it needs it, so it doesn't have to prioritize some functions and deprioritize or outright deprive others.

When the body is optimally resourced, it's like us when we've got financial freedom: We've got what we need when we need it and don't have to borrow from here to cover some shortfall over there.

Optimally resourced is your best shot. The way we get there is *optimization*.

CHAPTER 5

SH*T TO UNLEARN

The rise in poor metabolic health—which we now can more accurately describe as weight-health-ecosystem disturbances—has been met by an epidemic of solutions. One of them is to "know more" about nutrition, fitness, sleep, and genetics. The idea being that the amount of knowledge gained will directly correlate to an equal gain in health. The reality is that this solution has become a key contributor to a deepening epidemic.

Knowledge's partner in crime is for there to be a "right" answer served up as a superlative—the best, the perfect, and the ideal. As humans, we inhabit a body that never achieves a superlative, so when it comes to our weight health, no one number or result is ever truly static. Thus, while a superlative may sometimes be the best, it is almost never better. Which brings me to a favorite word: *better*. Recalling my origins as an advertising executive, my brain orients on how to sell. Years ago I realized I needed to sell the message of "Better, not perfect" to help patients embrace goals that are better for them and their bodies. Too many perceived failures—noting you can't fail if your goal was never physiologically possible—and way too much brain overwhelm occurs when we pursue superlatives in nutrition. Thus, as we move throughout this book, you will hear me use the term *better* in the form of recommendations or goal setting. "Better than what?" you may ask. Better than how your body currently feels. Better than what you observe your body is not responding to favorably. Better as a personalizable definition specific to you and your body today.

Before we move on, I have to call myself out. This amazing marketer had to listen to her better friends and surrender to their brilliance when it came to the title of this book. "Ashley, no one wants *their better shot*. The song is 'Hit Me with Your Best Shot'; the action is 'Give it your best shot.'" And here's where life goals differ from human nutrition recommendations. A superlative can be exactly what you need and want to go for in life. Now you have a book that will help you assemble several better choices that will result in your best shot at optimal weight health.

At the turn of the century, as I started my nutrition practice, I was keenly aware from my own experiences and my few patients to date that people were already INFOverweight. Between my background in marketing and my own health pursuit, I cued into two things:

- We can mistakenly believe that if we learn what is "healthier," we become healthier.
- Marketers craft messages based on what research indicates we crave and create a story to tell us how their product alone can deliver what we seek.

In 2005, while working in private practice, I also became a consultant for digital health companies. Those were the early days. I had an inside view into how patients would be able to access health information without practitioners, so marketers could tell them more stories, more often, more ostensibly tailored just to them.

That's how we became INFObese. Suddenly, rather than waiting for your practitioner to answer your question or give you a truly personalized recommendation, you could google it. On computers and handheld devices, rather than in books that (in theory, at least) had some research behind them, you could search for information to consider your own remedy or build your own health program.

Circa 2011, as a speaker at the first digital health conference of the Consumer Electronics Show, I heard the term *INFObesity*, meaning technology-driven info overwhelm and its negative consequences. The light bulb went off as I realized that the trend I was seeing in patients was dwarfed by what participants at this conference were building and launching (and financing) to "enable" consumers to curate their own data collection.

A year or two later I realized that we were already in the early stages of the INFObesity epidemic. How does a snowball become an avalanche? Everyone becomes an expert armed with data and mindsets; beliefs, not facts. And it went further, with platforms and vehicles making it possible for the armies of influencers to share their information with followers.

We are all clear that we can't healthfully handle the endless onslaught of information. Relevant to our work here, neither can our bodies—specifically our Switches.

INFOBESITY BLOCKS SWITCH OPTIMIZATION

The old axiom "Knowledge is power" needs a rethink. Specific insights about our bodies' needs and design, toward choices that will help us meet those, are empowering. We need tools to cut through all the other stuff, which at best is a distraction and in many cases is a disruption, to get to what's useful.

INFObesity misdirects you from achieving Switch Optimization as its many effects disrupt your entire weight-health ecosystem. Each of its parts—digestive, endocrine, nervous, cardiovascular, and so forth—experiences challenges as the body consumes excess information.

Multitasking and information overload can lead to decreased cognitive performance, and excessive information exposure significantly impairs our ability to recall and retain information.[1]

High levels of information overload can lead to increased stress, with detrimental implications for mental health. The strain of dealing with excessive information takes a toll on our well-being, leaving us anxious and overwhelmed.[2]

That is something we're increasingly conscious of. But what we may be less aware of is how the accumulation of pop science and (often poorly translated) research makes it darn near impossible for our vagus nerve, digestive tract, and ecosystem overall to connect and deliver what our brain and body need to run better.

---

1 Joseph Firth et al., "The 'Online Brain': How the Internet May Be Changing Our Cognition," *World Psychiatry* 18, no. 2 (May 2019): 119–29, www.ncbi.nlm.nih.gov/pmc/articles/PMC6502424/.
2 Miriam Arnold, Mascha Goldschmitt, and Thomas Rigotti, "Dealing with Information Overload: A Comprehensive Review," *Frontiers in Psychology* 14 (June 2023), https://doi.org/10.3389/fpsyg.2023.1122200.

That's the modern-day sh*t show—information overload about getting healthy is keeping us from becoming healthier.

I don't have national statistics on INFObesity, but my experience tells me that the figure tracks closely with the 93 percent of us whose metabolic health is suboptimal. In the absence of a clinical assessment, I've worked up a game: *Never Have I Ever . . . INFObesity Edition.*

NEVER HAVE I EVER . . .

☐ changed what I ate or drank based on a piece of information I'd taken in *on the same day,* from a source that had no knowledge of my medical history or my body's needs.

☐ noticed a change in my mood after interacting with a piece of information about someone else's nutrition or fitness choices and/or their image.

☐ panic-searched the internet because of a health concern, hoping to find a solution or image that matched my concern.

☐ failed to be fully present at a special occasion (vacation, meal, date, time with kids, coffee) because I was wrapped up in an internal dialogue about nutrition or health, based on something I'd just seen or read.

☐ spent money on something that sounded like it was a good solution for a health issue or route to a health goal, only to have that something fall short.

This is the start—whether you checked one or more boxes or none—to assessing how information is impacting your Switch. Later, in a follow-up assessment and recommendations, I'll cover how to pivot and reduce your INFOload by identifying your highest-value, "bull's-eye" information sources.

Here, right now, we can clear out the low-hanging fruit—mass information that's not helping anyone get or stay healthy.

WRONG STUFF IT'S TIME TO UNLEARN

- Total calories in a day: calories in and out.
- MyPlate is so much better than a pyramid, and always served with a glass of milk.
- Weight: The total number on the scale is all that matters.
- Blood sugar: To know if you're healthy, use a ninety-day average.
- Low cholesterol means you are heart healthy.
- Inflammation and stress are always bad; we should eat and act to prevent them.

These ideas are wrong for multiple reasons. For our purposes, the wrongest among them is their failure to recognize the body as a weight-health ecosystem.

The fix for all of them is recognizing suboptimal function and optimizing Switch function.

How brilliantly efficient! Yep, that's the body, by design. I'm obsessed with efficiency—for my patients, for myself, and for you.

Another win? As you unlearn, you'll be learning key points of the Switch Optimization plan.

You may be shouting, "Just lemme at the damn plan already!" This *is* the plan. We're in it. Unburdening your Switch of all this wrong info is key.

Let's start with the tired truism that's been feeding a trillion-dollar diet industry that thrives because it makes sure we never reach and maintain our goal.

1. Total Weight Doesn't Tell Us Who's Healthy

PERSON A   PERSON B: PERSON C:

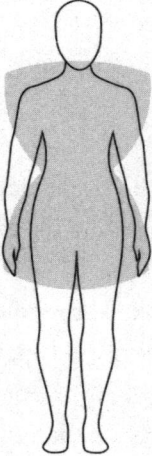 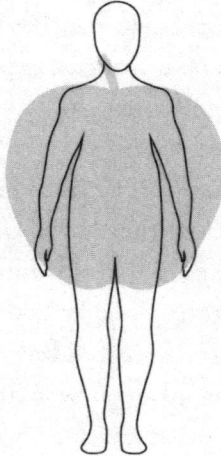 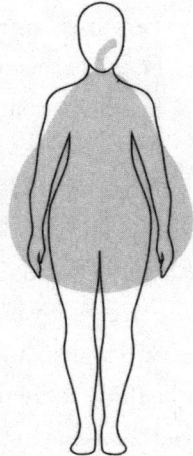

With a glance at the shapes above, we are programmed to think: Person A, healthy and in good shape. Persons B and C, obese or overweight and definitely unhealthy; they probably eat crap and don't exercise.

What these pictures don't show and what comprehensive assessment of these three individuals reveals is a wholly different truth:

Person A has unhealthy inflammation, suboptimal cholesterol and blood sugar levels, insufficient muscle mass, and unhealthy bone loss.

Person B has excess subcutaneous and visceral fat mass, good bone health, and lower body muscle mass, but needs more muscle mass in the upper body. They also have hormone imbalances, suboptimal digestion, and nutrient insufficiencies, notably vitamin D.

Person C has some excess subcutaneous fat, but visceral fat, bone health, and lean body mass are excellent; digestion and hydration could use some attention.

None of the three enjoys optimal weight health—right now. They're all part of the 93 percent of the US population who are, by a variety of measures, metabolically unhealthy.

Assessing weight health by weight alone overlooks a vast number of people who are close to some "ideal weight" and still weight unhealthy, while lumping all those who are overweight or obese together as having the same issue: just too big a number on the scale.

Weight *does* matter. It's a key performance indicator for health. It's just one among many. It doesn't matter exclusively, and not in the way we've put it on some kind of pedestal. Obsessing, as we have for decades, on that number on the scale is lame and misdirected.

When we're looking at weight health, optimization begins with *weight composition*.

WEIGHT COMPOSITION

Fat mass—type and amount

Skeletal muscle mass—muscles, bones, and their current status

Weight composition is the most important and actionable dataset for Switch Optimization. It reveals where things are better and what needs optimization. You have a weight-health ecosystem—the Switch is the controller. The Switch set of hormones helps regulate weight composition—the proportions and locations of fat, bone, and muscle. Thus, assessment and optimization of Switch function is core to upgrading your weight-health ecosystem.

That's one big *unlearn* out of the way. Let's knock through the others on our starting list:

Mass nutrition and fitness guidelines are all good and interchangeable.

What you take in gets delivered and used.

Average low blood sugar means healthy.

Low cholesterol equals heart healthy.

2. Calories, Pyramids, and MyPlate—Oh No!

As we progress in our unlearning, we arrive at my favorite lesson. I know I am not supposed to pick favorites, but if I had a dime for each of the diet "failures" and the resulting products, guidelines, and programs this sh*t has generated, I would be on my own island with more coconuts than I could ever need.

About a decade ago, having just been selected as the California state representative for a national health campaign, I uprooted myself from Los Angeles and moved to Washington, DC, because I believed everything was signaling a new era in government to improve health *powered by better nutrition*. In DC, I had a seat at several tables. One of them was at a pre-awards dinner with executives of global food companies. Ironically, what prepared me for what unfolded was not my career as a dietitian but my earlier career as an ad executive. By way of openers, each person introduced their company's product improvements and results for the table to applaud. The first was the reduction of calories in yogurt—by the removal of fat. The second was the reduction of added sugar in yogurt—with the addition of nonnutritive sweeteners. The third was the increase in fiber in snack bars by adding inulin—while the remaining ingredients were highly processed and poor quality. And finally, the soda company revealed its win—selling the same sodas but in smaller cans.

Around the table, heads were nodding in approval.

You and me, right now our heads are exploding. We know this is BS . . . today. Back then, I took one for the team. I didn't hold back: Are you kidding me? Shrinking sodas? Removing the fat from yogurt reduces calories, but it changes the way the yogurt impacts the body, and not for the better. We know you can't cheat with nonnutritive sweeteners without the body keeping score. And adding fiber to a nutrient-deficient, even problematic product doesn't qualify it as a health win. Hey guys, why aren't we focused on better nutrition for better health?

The looks on their faces or their failure to look at mine told me all I needed to know. I didn't belong there. I was at the wrong table.

I was elated. I went home. The next day I skipped the awards. Instead I wrote what follows, and over the next few months I started my company:

> Better nutrition gives your body what it needs to run better, today, while reducing or avoiding what irritates, overwhelms, and disrupts those efforts.

If you're thinking, "Oh wow, it was so bad back then," hold on. The same thing is still happening: Seed oils will kill you, beef tallow forever. Remove the red dye but not the loads of added sugar. High protein but modified fat! The list sadly goes on.

Let's keep going with our sh*t list.

- Anything that is one-sized. All men or women need X. People of a certain age or specific dietary pattern need or should have this. Every meal should have this and not that. A plate should always be accompanied by a glass of milk. A meal is made up of . . . , whereas a snack can include . . . and so on.

- Calories go in and calories go out. The end. That alone is not a plan for optimal health. To be optimally resourced, the quality of your calories, their nutrient balance and quantity at one time, the timing when you start and stop having them, how your body can break them down and use them today—that's what matters . . . more.

- The food pyramid (or a plate) is a better guide for you. Pyramids belong in Egypt, where we can speculate about how they were made and stand in awe of those who created them. A pyramid food picture is at best not useful for driving different food choices. It can also contribute unfavorably to imbalances in an individual's nutrition. We can and should experiment, assess, and curate your nutrition plan in response to what your body needs and the choices that are deliciously doable for you today.

- Supplements and medications for the win. These are tools, not total health solutions. No supplement or medication can optimally resource a body, and each impacts the body's ecosystem. Their use must be considered in coordination with your initial assessment and reassessment to determine their ongoing value and impacts.

3. More Isn't Always Better

When I was in school for nutrition, I was taught that when in doubt, the answer to any question is "small frequent meals." It actually works a lot of the time. We aren't here to learn or unlearn that one. We are here because *more* of anything—on its own, with no context—is not always a better nutrition or lifestyle recommendation. Our goal is to be optimal, and we optimize resources by finding the better amount your body needs today. Yes, that could be *more* than you are currently consuming or doing, but how much more, what type of more, your body's response to more of one thing, and the overall effect on the ecosystem are crucial to identify. More protein, more calcium, more fat can mean more problems, especially in an ecosystem with suboptimal digestive function. More strength training for a body that is stressed, injured, imbalanced, or already at its max needs will likewise disturb or increase risks in the ecosystem. Likewise, if someone who is not doing or is doing minimal strength training right now hears the "more" message and adds a daily plank or some light weights for a few minutes, their "more" may be better but not enough to be optimal. Strength training is exercise that asks the body to exert effort against added resistance. It does loads for the body, and your body needs it for diverse reasons, one of them being optimal function of your weight-health ecosystem. The added resistance can come in various forms and at varying times of the day.

I got schooled on this during my first trip to Alaska to go fly-fishing. I was psyched. I was about to catch a floatplane to a tiny island in Bristol Bay. The guy checking me in for the flight weighed my bag. Then he weighed me. Together, we had to come in under two hundred pounds. The next passenger—another Hollywood person, but unlike me, a famous one—had a bag, a cooler big enough to host a tailgate party, and a duffel that was bigger yet. The guy could barely heave the duffel onto the scale. How was this going to work out? Oh, a separate floatplane for the celebs. Right.

The cooler, it turned out, was full of meat. Out on the island, our guides had stocked some of the best game in the US. And here was meat imported from California. Okay, then. The remaining mystery was that duffel. The guides got a workout hauling it around. Its contents, we eventually discovered, were dumbbells, kettlebells, barbells, and hell's bells—enough iron for a Gold's Gym. Somebody was definitely going to stay in shape. Hat's off. The hilarious thing was that the guides did more resistance training on that trip than the duffel's owner. And they didn't need any gym weights. They got all the heavy lifting they needed being guides and lifting bags, anchors, and more.

Takeaway: None, really. Just that life is weird. If there is one, it's this: We can strength train in all kinds of ways—in a gym, at home picking up kids, on a farm, at Costco. It's more than simply "do more," but it isn't necessarily about specialized equipment. The core thing (pun intended?) is making sure our fast- and slow-twitch muscles work against enough resistance to maintain or build lean mass. That happens through resistance and stress that compel shifts in consumption and elimination (via waste) of molecules. The appropriate, effective resistance for you—the right form and the right amount—depends on the resistance you get in day-to-day life, your body's current health status, and your body's needs.

Got it? It's personal. To optimize it, you should personalize it.

I'm not a fitness expert, so I partner with people who are. When I do, I give my patients some experiments. The aim is to gain insights into what their bodies need, what they'll respond to, and how they're responding as we roll. In building your plan we'll do the same.

Seldom does personalizing mean we have to burden guides in Alaska with hundreds of pounds of weights in our duffel bags.

4. The Pitfalls of Average Blood Sugar

The Shot continues to reveal issues with a checklist-system approach to health care. Blood sugar provides yet another example of how failure to assess and treat ecosystems produces suboptimal outcomes. Recall that the Shot was developed to address an epidemic of diabetes—a disease caused when sugar isn't operating properly in the body—and suboptimal treatments. Weight loss was a side effect. And that made the Shot an unignorable fashion statement.

This shows us the ecosystem at work. Blood sugar doesn't operate in isolation from the rest of the body. It is a core part of weight health, sharing the same hormones that signal appetite and fat accumulation in response to consumption of carbohydrates, protein, and fats. Suboptimal Switch function will result from and result in suboptimal blood sugar levels. But the way we conventionally discuss, assess, and diagnose elevated blood sugar doesn't show the whole picture. In fact, it's just about as problematic as the problem itself.

When a patient comes in with a blood sugar issue, it's common practice to look at their intake of carbs—what kind and how many—and evaluate their blood sugar as a singular factor. That's wrong. Even more wrong is how blood sugar is only investigated with regard to diabetes. Blood sugar optimization is a foundational piece of weight-health-ecosystem operations and overall health. Thus, optimization, regardless of your levels, is a requirement for everyone, beginning with assessment in a *better* way. That's what you are here to do, recognize that you want optimal health, which requires the body to be optimally resourced.

For a long time, we just measured fasting blood sugar, which helped us see in-the-moment dysfunction, but nothing else. "Fasting" in this context means no food (except black coffee or tea) for about eight hours, so for most people, we take that measurement first thing in the morning. But it's hard to get a blood draw *immediately* when you wake up. Patients need to wake up, deal with getting their kids out the door and the dog walked without any fuel, make their way to the lab or the doctor's office, and then do the blood draw—by which point there is nothing ordinary or normal about this day for having their "fasting" blood sugar checked. Or "fasting" might mean eating in the morning and then eating nothing for eight hours, for the test. That is not representative either. The result is one's fasting blood sugar level isn't indicative of a "normal" fasting level.

Recognizing that there are forty factors that impact our blood sugar level, and that only ten of them are nutrition-driven, it's easy to see how data collection without this consideration can produce skewed results. A big non-nutrition factor is stress, particularly the stress of asking our body to do a bunch of stuff when it's been deprived of fuel. A nutritional factor is caffeine, the response to which is individual (genetics) and can include what we add to that black coffee (such as nonnutritive sweeteners, the sugar cheat we hope our body doesn't decode). For these reasons and more, fasting glucose just is not that helpful a number, not even in the diagnosis of type 2 diabetes, in which it misses many millions who don't present with in-the-moment elevated blood sugar.

Here are forty factors that affect blood sugar levels. Note that only the last ten—bolded—are nutrition or food related:

- Elevated stress
- Insufficient/disturbed sleep
- Time spent sedentary
- Time spent moving
- Exercise type
- Exercise intensity
- Recovery (post exercise)
- Sexual intercourse/ masturbation
- Digestion
- Inflammation
- Hormone fluctuation
- Genomics/genetics
- Illness
- Injury
- Medications/dose
- Nutrient interactions
- Medication interactions
- Pain
- Puberty/menopause/ pregnancy and postnatal
- Period (menstruation)
- Insulin sensitivity/ resistance
- Scar tissue/lipodystrophy
- Temperature
- Altitude
- Trauma(s)
- Perceived lack of control
- Decision fatigue
- Anxiety/fear
- Connection/disconnection
- Time in nature

- Food/supplement quality
- Food/supplement quantity
- Food/supplement nutrient balance
- Food/supplement timing
- Caffeine
- Sweeteners
- Hydration
- Alcohol
- Intolerances/allergies
- Nutrient deficiencies/ insufficiencies

In an effort to address the flaws of fasting blood sugar for diagnosis, enter average blood sugar assessment: hemoglobin A1C. At the time of its debut in 1968, A1C was a significant upgrade in blood sugar evaluation and diagnosis. With this tool we can see, over a period of ninety days, how much glucose is sticking (glycation) to hemoglobin in the blood. With the ability to get some insights about three-month activity—non-fasted, actually lived—A1C became the gold standard for diabetes and prediabetes diagnosis. An improvement. Incremental, but a game changer.

One problem with any average is that all kinds of readings—an infinite range—can take us to the same figure. Last time I got pulled over for speeding, I tried this: "Sure, Officer. True. But my *average* speed was fifty-five." He wasn't buying it either.

The following chart shows three people with an A1C of 7. The dark gray is "time in range"—range being the desirable level for their blood sugar. (Ignore the range numbers. Focus on the line running through the range that represents their blood sugar.) While person A experiences rolling hills in range as a response to choices, persons B and C are both outside the range, from a little to a lot.

A1C, while much better than fasting blood sugar, isn't an optimal dataset.

To optimize blood sugar, we need better data.

PERSON A:

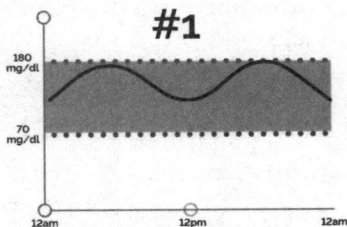

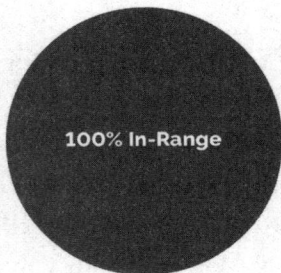

PERSON B:

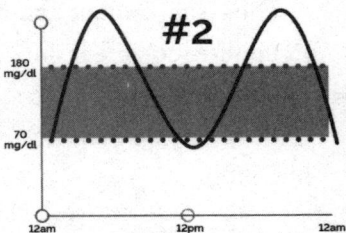

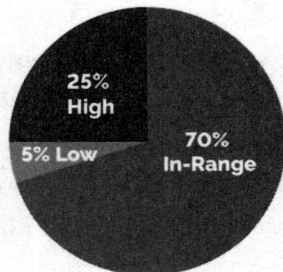

PERSON C:

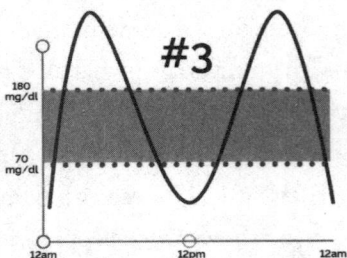

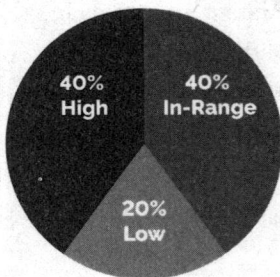

How do we get the intel needed to optimize blood sugar?

To assess *your* blood sugar function so we can optimize it, we need to know not just an average level but how the level changes over time. Specifically, these things:

How your current choices are impacting your blood sugar

What drives your blood sugar "out of range"

How high or low your blood sugar goes

How long your blood sugar stays there

Meet Lou. At the start of our work together, he wore a continuous glucose monitor (CGM) sensor. He followed his current routine and choices, tracked them, and got real-time feedback. After five days, I reviewed and shared some suggestions—experiments— for the next week. He implemented several. The results on the next page show a 54 percent improvement in time in range, a 23 percent decrease in average blood sugar, and a 48 percent reduction in glucose spikes. Lou went from very suboptimal blood sugar to nearly optimal (more than 90 percent time in optimal range for a person without diabetes, and about 100 for average glucose). The total time for this dramatic weight-health improvement was two weeks. The total cost: about $299 (subtract $99 if insurance covers his sensor). In Lou's estimation, prior to this, he had spent several months and about $800 on different efforts—namely, supplements and specialty food products—to improve his blood sugar. His doctor wanted to put him on a blood sugar medication but was open to Lou trying dietary changes first. Had Lou adjusted his diet without a continuous glucose monitor, he would be like the millions

**LOU'S FOURTEEN-DAY CONTINUOUS GLUCOSE
MONITORING RESULTS**

	TIME IN RANGE	AVERAGE GLUCOSE	SPIKES/DIPS
WEEK 1	31%	140	25
WEEK 2	85%	108	13

who typically just give up carbs, which proves unsustainable and leads to or exacerbates existing ecosystem disturbances. With the CGM, one of his biggest aha moments occurred when he added almond butter to his pre-walk banana (a carb), adding protein and fat to slow the absorption of the sugar. He thought he'd have to give up the banana based on his initial readings, but he liked the taste, potassium, and energy for his walk.

Continuous glucose monitoring gives us that. A wearable sensor assesses blood sugar virtually in the moment by tracking interstitial fluid, which gives a reading a bit different from a finger prick. This shows both the trend and cumulative blood sugar responses to nutrition choices and to the rest of the forty factors. CGM gives us better data on your body's needs and how close to optimally you're resourcing to meet them.

Additionally, we can now measure how much insulin your body has available when fasting. Fasting insulin enables better diagnosis of impaired function of this key peptide hormone. When evaluated along with CGM data, it can help further personalize recommendations.

Because Switch hormones GLP-1 and GIP incite insulin into action, fasting insulin also gives us a window into suboptimal

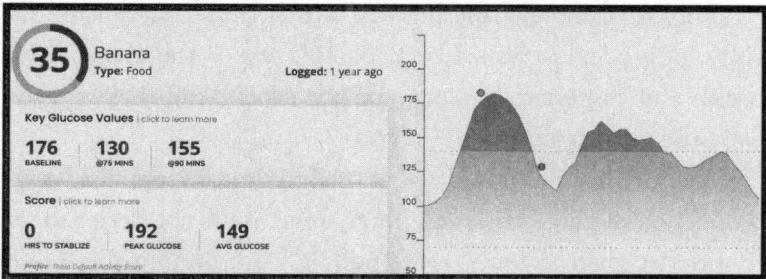

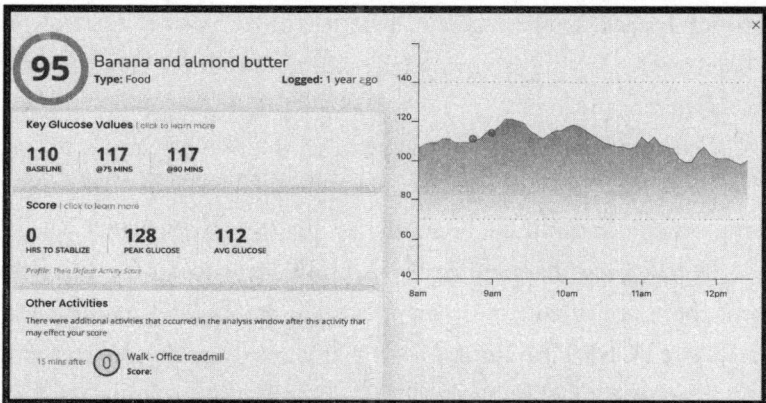

©Theia Health

Switch function. Our bodies deploy insulin in response to varying levels of blood sugar. Our ability today to assess not just its levels but what is impacting its function has a profound impact on our optimization mission. It makes the mission much more doable, capable of being personalized, and precise.

5. At Least My Low Cholesterol Says I'm Heart Healthy!

Sorry, Charlie.

Low or "normal" cholesterol does not by itself mean "heart healthy." Ever.

I'm hell-bent on this one. A zillion-dollar faction of the INFObesity industry has been putting it out there that heart health and cholesterol are one and the same. And they come in one little pill: statins.

These drugs are formulated to run interference with the body's production of cholesterol and lower some numbers. Better to say they're designed to lower some numbers by running that interference. So that number has been put on billboards and tattooed on our brains. Lowering it can be a win to reduce "traffic" on cholesterol-carrying highways. Less traffic reduces risk of some accidents. Some.

Overriding the body's cholesterol production has implications. Here's a short list: lower testosterone; lower coenzyme Q10, a potent antioxidant; muscle breakdown; digestive disturbance.

When these happen, or if the body dares to ignore the first pill, there are of course other pills and creams, and medications such as a PCSK9 inhibitor ("inhibit" means override). With them the liver gets rid of cholesterol. There are work-arounds, such as supplemental coenzyme Q10, electrolytes, and probiotics. And when your testosterone drops, there are pills to deal with erectile dysfunction—from the same companies that sell us the statins—and hormone replacement and peptide therapies.

Because the priority is lower cholesterol, right?

Nope.

Optimal heart health means that the heart, in receipt of regular signals, works effectively and efficiently. Suboptimal heart health occurs when the heart doesn't receive signals. Maybe they don't arrive in time or at all. Maybe they're in a language the heart doesn't understand. Maybe there's too much traffic or too little space for travel due to disturbances in the lining of the freeways. Maybe there's injury to the system's parts.

Cholesterol is *one* part of the heart-health picture. The type of cholesterol we typically measure—just like the type of weight we typically measure—gives us little helpful data, even inaccurate data.

We have some anecdotal experience about this. Who doesn't have a friend, colleague, or relative with "normal" cholesterol who had a heart attack, a stroke, or a near miss?

To optimize heart health, we need a whole-person approach that looks at calcium, plaque (hard and soft), endothelial glycocalyx, antioxidants, blood pressure, potassium, elimination, fiber . . . and, yes, genetics. But not just the "my dad and his dad had high cholesterol" genetics. I'm talking the gene stories that answer questions like: How are you uniquely designed? Do you process B vitamins such as folate, B12, and B6 efficiently to reduce homocysteine production, and if not, do you need different choices or additional methylation support?

6. You Are What You Eat

We all know the myths "if you drink water, you become hydrated" and "if you eat kale, olive oil, and wild salmon, you are healthy" and "if you go on a detox, your body detoxifies."

A long-perpetuated myth and slogan is that we are what we eat. It's just not true.

Our ecosystem does not function optimally when we are in receipt of items deemed healthy.

It runs better when it recognizes, signals, breaks down,

absorbs, uses, and discards what we bring it efficiently and effectively. The myth of taking it in as the only thing that matters has created so many problems. At the core, misunderstanding hydration and digestion.

Dr. Anup Kanodia summed up all my conundrums about existing hydration recommendations—their impersonalized nature, the conflicts, the inaccuracies—when he suggested that I ask patients, "Are you a hose or a sponge?"

My dad, a urologist, taught me that a good way to look at hydration was the hue of your pee. Drinking water—eight 8-ounce glasses per day—doesn't guarantee that you're hydrated. You are not necessarily *better* hydrated if you take your body's weight in pounds, divide by two, then drink that number of ounces of water each day. Nor will chugging even more water ensure that you're *optimally* hydrated. Nor does it for sure mean you're well hydrated if your urine is clear. Ditto adding one or more electrolytes supplements to your water each day. Or peeing all day. Or never being thirsty. Or counting all the water in your fruits and vegetables, coffee and tea . . .

You *are* optimally hydrated if you are a sponge and not a hose.

Hydration is a process that goes way beyond the body's intake of water. It involves ensuring that an optimal resourcing of water gets to multiple places so that it can do its many jobs. Water may combine with fiber, expand it, and pick up waste. Water may also pick up water-soluble nutrients and compounds; optimal levels of other nutrients and hormone signalers permitting, it then delivers those nutrients and compounds into cells for metabolic functions. Some water hangs out in the bloodstream to support a more fluid environment. This is just a short list of the tasks water gets involved in.

Hydration is one of the weight-health ecosystem's core processes. Switch hormones play a key role in hydration status, and hydration is a key trigger for Switch functions. Hydration is critical to ecosystem function, and the Switch has a role in its regulation.

A sponge takes on water. Its capacity enables it to expand and hold even more. The sponge, too, takes on a variety of jobs: absorbing water in one place and transporting it to another; wiping up waste, taking it elsewhere, and getting rid of it.

A hose is made for water to flow through without much effort. The water goes in one end and out the other. That's all. That's what the hose is for. The water's headed somewhere else. It doesn't stick around.

So, which are you at the moment—a hose or sponge?

Digestion follows the same course. You need to be a sponge, not a hose. But digestion has more stages, so we will get to that assessment on page 110.

PULSE CHECK

How's this feeling so far?

Offloading INFObesity can mean feeling lighter and better. It can mean feeling a bit overwhelmed in a different way. Sometimes it works out to both.

These long-cherished weight and nutrition truisms can be hard to let go of, despite all the evidence they haven't been working.

We're not simply ditching them, though. We're swapping them out to enable us to put in place better, more actionable data and science with a realer relation to weight health.

If that feels right, then we're ready to put our mission on a vector.

CHAPTER 6

PRIMARY SWITCH OPERATIONS

The ecosystem involves hundreds of hormones and peptide chains. The Switch has four main drivers, covered in the chart on the next page.

The Switch's basic jobs are to:

- stimulate release of **fluids** that break down food into usable nutrients;
- set the pace for food to pass through **digestion** for optimal **absorption**;
- trigger other hormones to act in the **brain**, letting the body know it is in receipt of what it can expect to be useful resources;
- ready the **heart** muscle to receive messages telling the heart that it has useful resources coming in;
- wake up and deploy **insulin**, the sugar-seeking hormone, to navigate blood, pick up sugar, and deliver it to cells for use; and
- optimize **hydration** by contributing to the amount of sodium eliminated via urine, sending messages to the brain to trigger

thirst, assessing dehydration in the colon, and adjusting water and electrolyte levels accordingly (to improve cellular hydration; this can lead to dried-out stool that is harder to pass).

These functions are carried out by the big-four Switch hormones, working in collaboration.

Here are those Switch hormones' individual tasks, lined up with their names:

GLP-1	
GLUCAGON-LIKE PEPTIDE-1	
WHAT IT DOES	HOW IT WORKS
Switches on insulin and leptin. Down-switches glucagon and gastric emptying, which helps set the pace of digestion and glycemic response. Switches off bone formation. May impact fat type, distribution, and metabolism.	Secreted from intestinal mucosal cells on receipt of nutrient stimuli (carbs, protein, fats). Vagus nerve fibers in the stomach send messages that also trigger release. Travels via bloodstream and vagus nerve to receptors.

PYY	
PEPTIDE YY	
WHAT IT DOES	HOW IT WORKS
Helps set the pace of digestion. Helps switch on satiety response (feel full). Increases water and electrolyte absorption.	Gets messages regarding arrival of calories, especially calories in protein and fat (amino and fatty acids) lower in the small intestine. Exercise can increase fasting levels. Caloric load and medications impact the amount secreted. Travels via bloodstream to receptors.

CCK	
CHOLECYSTOKININ	
WHAT IT DOES	HOW IT WORKS
Switches on gallbladder and pancreas to release digestive fluids (bile, enzymes) for the breakdown of food into nutrients.	Released when amino and fatty acids arrive in the small intestine. Stomach acid (hydrochloric acid, HCl) levels and acetylcholine (from vagus nerve) impact secretion. Travels via bloodstream and vagus nerve to receptors.

GIP	
GLUCOSE-DEPENDENT INSULINOTROPIC POLYPEPTIDE	
WHAT IT DOES	HOW IT WORKS
Switches on release of insulin and glucagon from pancreas. Switches on bone formation. Impacts fat metabolism and inflammation.	Released when glucose and amino and long-chain fatty acids arrive in the small intestine. Travels via bloodstream to receptors.

That's where we are today. And plenty TBD. We keep learning things about the Switch and the weight-health ecosystem including the above.

THE SHOT IS NOT THE SWITCH; HERE'S HOW THEY'RE RELATED

The Shot, a game changer in weight loss, looks a lot like the Switch: GLP-1RAs, GLP-1 agonists, and GLP-1/GIP dual and forthcoming triple plus agonists, marketed under brand names such as Wegovy and Ozempic, and delivered in pill form as well as via injection. The Shot is hormone replacement therapy, substituting for the Switch hormones. It has provided benefits to millions, from better blood sugar to weight loss to reduced inflammation to even relief from sleep apnea. Its discovery, its mass adoption, its trials and tribulations are something bigger than what it gets attention for most often.

The success and challenges of the Shot reveal an epidemic of suboptimal Switch function.

The Shot's achievement is making us aware that our weight-health ecosystem and Switch exist, and that for most of us, they could really use optimization. It's because we have the Switch that science could invent the Shot, and now the Shot is helping bring the Switch into the conversation.

Whatever our relationship to the Shot, it will go down in history as the tool that taught us about and allowed for deeper investigation into Switch function, and helped us—me first, now you—devise a plan to optimize it now and going forward.

The big reveal within the reveal: Modern-day existence throws a whole lot at our Switches, so much that we have an epidemic of suboptimal Switch function. This epidemic makes Switch-hormone replacement therapy a helpful or even necessary intervention for many of us. And it makes Switch assessment and optimization, as indicated, mandatory for all.

The Switch-Shot relationship hinges on turning on and turning off.

Our natural Switch hormones turn on as needed, as indicated by various inputs, not only nutrients; and they shut off pretty fast, within five minutes. The weight-health ecosystem sends out enzymes that seek and break down the Switch hormones. It's time for rest-and-digest mode.

But how we live has changed, and rapid shutoff has some downsides.

The Shot's synthetic hormones were engineered not to shut off in minutes. Or in hours or even a day. They stay on for about a week. In the future, likely a month, possibly longer.

Go-on-and-stay-on is the game changer. It's what's made the Shot heard round the world.

What's the trouble with quick shutoff? After all, it's how the Switch is designed.

Let's think of the Switch as a motion detector. A motion detector waits, alert but not active, for a person, an animal, a robot, or any moving thing to pass. The device receives data saying "motion" in its surroundings. It sends a message via an electronic signal. That message triggers multiple actions:

A sound beeps or blips or rings and/or a light goes on.

A person in range of the sound or the light looks at a monitor.

Or no one reacts on site.

The device sends an alarm to the police or a security company.

By the time the *re*actions are happening, the initial action—motion—is often over. There's no longer motion out there to detect, so motion detection stops. Yet the light stays on, the beep's still beeping, and maybe the police are on their way.

The Switch detects food. It sends the message: "Food." This tells our body to be ready—to receive nutrients and for work to be done. The Switch sends those messages, and it's done its job. It shuts off.

That's how the Switch was set up to work. In the highly regular and constant world we inhabited until the industrial revolution, our weight-health ecosystem's messaging was in perfect sync. In that world, we hunted and caught our food, and in due course mainly grew it; we slept and woke mostly by the rhythms of natural light; the messages we sent and received were few, specific, and focused.

Fast-forward to today. Picture our motion detector—the Switch—with twenty-seven dogs, three motorcycles, a trio of joggers, a drone, and a runaway hippo all charging at it at once.

How can the motion detector send an actionable message? "All hell's breaking loose" is not much of an actionable message.

Modern living is subjecting our Switches to that kind of chaos.

There is nothing more chaotic than the bombardment we subject ourselves to—and so our ecosystems—in the form of sugars and sweeteners. From recommendations telling us to avoid all sugar to those saying replace sweeteners with nonnutritive ones, on or off the Shot, we still impair our Switches. The results? Irresistible cravings, belly weight, mood shifts, and energy and blood sugar challenges that so many of us battle daily.

So here and now, let's set the record straight on what isn't and is better, as the sweet stuff goes.

Sugar—being the basis of one of the three macronutrients, carbohydrate—is the primary fuel the body is designed to run on. So optimally resourcing the ecosystem naturally involves some nutrition in the form of sugar.

That's a horse of a different color, though, from the inundation of added sugars and nonnutritive sweeteners coming at us in food products, beverages, and supplements that are engineered to cultivate a codependent relationship with the consumer.

That relationship damages our Switches and ecosystems in

various ways, covered throughout this book. While you have likely heard all you ever want to and more about sugar's negative impact on your body and its systems, you haven't heard about its impact on your ecosystem and the Switch, its regulator.

Here I will provide the CliffsNotes version and your first experiment. Later I will go into more specifics.

Digestion + Hydration + Better Nutrition →
Switch Function → *Ecosystem Optimization*

"I am drinking the iced tea that's 'just a tad sweet, and my blood sugar still goes up?'"

"I gave up sugar and only use Brand Z nonnutritive sweetener—but I haven't lost any weight."

"Ashley, can you help us with this article on which sweetener is better for anyone trying to lose weight?"

Those are just the tip of the tip of the iceberg.

Confronting nonpersonalized recommendations that fail my patients, my MO is to devise experiments that show them why off-the-rack recommendations are unlikely to fit them and then to give them something that can specifically reveal what their bodies need to run better.

At a point, I realized I had one too many patients consuming more added sugar than was better for their bodies at the same time that I had others who were doing the same but with nonnutritive ones. I needed a way to show them both that what matters more isn't a set amount of either or that one was better than another. What matters is something more physiological: how our sweet taste buds are working and the information they're sending to the rest of the ecosystem. Thus, assessment of sweet taste bud function provides us a key performance indicator of Switch and ecosystem function.

That's how this test came together.

To start your Switch Optimization, take it now.

THE SWEET TASTE BUD TEST (AND RESET)

You will need a few bites of an apple (a quarter or two slices). With the exception of a green Granny Smith or candied or baked apple—which are not allowed—the type does not matter that much. If you can choose one that isn't advertised for being super sweet, that is better. It is more important to use the same apple type—and about the same degree of ripeness—so your test and retest get accurate results. If you cannot consume apples, you can use a pear instead.

Consider previous bites and sips that are sweet to you and mentally develop a scale—from 1 to 10—where 1 is not sweet at all and 10 is very sweet, the most sweet that you can recall.

With a neutral palate (not following recent toothbrushing, using mouthwash, or consuming coffee, alcohol, food, gum, or mints; if you have, you can eat a few slices of cucumber or sip on some water), take two to three bites of your apple, chewing them thoroughly.

As you chew and experience the flavor, score your bites on your scale from 1 to 10.

On average, what do your bites score?

Results

- If your apple bites are greater than a 7, then your sweet taste buds appear to be well-aligned with the natural degree of sweetness.

- If they are a 6 or less, you will want to conduct a reset and retest your apple bites every seven to ten days until you achieve a score of 7 or greater.

The Sweet Taste Bud Reset

Remember that this is for a set period of time, not a life sentence.

Remove all sugars and sweeteners—sugars, syrups, honey, and so forth—that you add to your food and beverages. Remove all ready-made products containing them. This includes all non-nutritive sweeteners . . . yes, even the natural ones. If your current supplements—including fiber and protein powders—contain sweeteners or added sugar, work with your practitioner or make replacement choices for the duration of the reset. You may be onto something with your sweet taste buds if you discover these as a source!

Yes, you can have foods that have naturally occurring sugars such as fruits and vegetables— incorporating them into your overall nutrition plan may really help you make it through the reset! Baking or preparing fruits and vegetables with a pinch of salt or spices such as cinnamon may help bring out their sweetness to help satisfy you. Herbal teas that have a natural sweetness can be consumed as well.

During your reset, document and experiment with strategies to address when you experience sweet cravings.

Are you tired? Hungry? Thirsty? Emotionally high or low? Did you consume something high in salt? Did you have a meal or snack that had protein and fat but no carbohydrates? Was what you ate or drank not delicious to you in the moment? Maybe a sweet craving hit post caffeine or alcohol? Any signals from your digestion—gassy, bloated, constipated, or loose stools? These insights will help you optimize your sweet taste buds and your ecosystem using the discussion points and additional experiments in the book.

Redo the Sweet Taste Bud Test after seven days and continue the reset protocol if you are not yet a 7. When you achieve a 7, you can experiment with bringing back choices such as added sugar and sweeteners. Informed by your insights, experiments, and the reset experience, the amounts that you add back should help you maintain optimal sweet taste bud function.

Retest once a quarter or when you notice you are challenged by sweet cravings.

Here's an overview of some of the additional problems in which Switch issues and suboptimal Switch function are implicated:

SWITCH ISSUES AND SUBOPTIMAL FUNCTION

Diseases and syndromes: Crohn's disease, ulcerative colitis, celiac, Hashimoto's disease, Graves' disease, scleroderma, rheumatoid arthritis, psoriasis, psoriatic arthritis, prediabetes, diabetes (gestational, type 1, type 2), polycystic ovarian syndrome (PCOS), metabolic syndrome, obesity, cancer, multiple sclerosis, epilepsy, gout, irritable bowel syndrome (IBS)

Disorders, conditions, exposures (or symptoms): leaky gut/intestinal permeability, gas, bloating, constipation, reflux/GERD, gastritis, gastroparesis, high cholesterol, high blood pressure, migraines/frequent headaches, small intestinal bacterial overgrowth (SIBO), H. pylori/ulcer, diverticulosis/ -itis, chronic fatigue, chronic pain (joint, back, overall body), fibromyalgia, premenstrual cramps or disorder, sleep apnea, low testosterone/estrogen/progesterone, adrenal insufficiency, elevated cortisol, insulin or leptin resistance, depression, anxiety, seasonal allergies, histamine intolerance, mold exposure, elevated heavy metals and toxins (glyphosate, etc.), food allergies or intolerances

As if to make things harder for our Switches and weight-health ecosystems, we've cooked up "solutions" for nonproblems and dreamed up products to satisfy invented needs. We've made conveniences, desires, and whims necessities. And, of course, we've repurposed sedentary living as leisure and a luxury. Which is why we need more than a motion detector.

These days, we need a force field.

A force field wards off attacks and intrusions. Better than a motion detector, which just alerts us to things that aren't stationary, a force field fends off inputs and stimuli that cause interference. This protective barrier will help you avoid the onslaught of unrecognizable, confused, inaccurate, even batshit-crazy information—all day, every day; at the grocery store, on social media, while dining out, and even while you sleep.

The Shot can put up a temporary force field that deflects a variety of things, mainly calories and food noise—and deflects them *hard*. The force field we'll build in optimizing Switch function does those things and more. It stays on as long as you continue to optimize, and it supports the whole ecosystem—a true fix that optimally resources your body.

Now our mission vector looks like this:

Plan → *Assessment* → *Optimization* →
Your Force Field = Your Best Shot

The medical, nutrition, and health-care industries, and armies of influencers, have dished up countless plans. They're in business to keep coming up with more. The trouble with most plans? Not that they go wrong so much as that they go nowhere. Or they go in circles.

Go keto.

Did it work?

No.

Okay, try vegan.

Did it work?

No.

Try fasting . . .

Try a statin.

Did those work?

No.

So try Repatha . . .

As Jay Z sings, "On to the next one."

It's a cascade. And the basic nature of a cascade? It goes downhill.

All these plans are different, and yet these plans are all the same in at least one way: They're all sold as one size fits all.

We need more than a plan. We need a mission. I learned the difference from *The A-Team*. As a kid, I was hooked on it. A bunch of unlikely military pros impressively and hilariously take on the most incredible assignments and get the job done with flair and exaggerated drama. Based on one set of mission specs, the A-Team determines the right tactic is to night-parachute into the jungle and catch the bad guys by complete surprise, free the hostages, and air-extract out.

All goes per plan. Until a rocket-propelled grenade hits the plane, forcing an emergency landing in a clearing peppered with landmines. Or until it turns out, mid-flight, that their pilot is on the bad guys' payroll.

So, change tactics?

No way. We're 2,500 miles from base, over territory dense with hostiles! The answer is to modify and tweak and torque *this* plan. The one we're in the middle of executing.

The A-Team's commander, John "Hannibal" Smith, portrayed to deadpan perfection by George Peppard, supplied me with one of my main mantras. Once the A-Team secured the episode's win, Smith reappeared, sitting or leaning back with his trademark cigar and grin. "I love it," he'd say, "when a plan comes together."

When a plan comes together, it's a mission.

Thwarting terror attacks and rescuing hostages held in steamy jungles may not seem to have much to do with our Switch Optimization mission, but stick with me.

Here's another thing we can learn from the A-Team: Do a situational assessment. The assessment tells us the mission's actions, the order, and what tools the mission calls for. Set out on your mission and do a situational reassessment to ensure your plan comes together.

Mission-critical equipment is key. When a force field is what you need, a motion detector isn't gonna cut it.

PART III

THE SWITCH OPTIMIZATION PLAN

Here's how we optimize primary Switch operations.

Personalization is what we want and need. As we move into the plan, let's look at this term more in depth to help you see how you will use your plan for Switch *and* ecosystem optimization.

Personalization gives a person's body what it needs to run better. You are unique. Your body's needs and the choices that you find deliciously doable are different from those for my body or what another person will find accessible and enjoyable. I borrow a term from the psychotherapist and author of *Fortytude*, Sarah Brokaw, who's also a longtime friend: "Research is me-search." As we move through this plan, therefore, you are not just invited but *expected* to determine which choices you want to experiment with. That will be influenced by what feels doable and the guidance shared regarding a better order or what should not be selected based on certain considerations. In the plan building and assessment that start here, no less than half of what you'll consider lies in the area of me-search. It's how you work—your body, mind, personality—and it's how different choices work for you. By experimenting, you'll zero in on better choices and also discover ways to check in with your body, get data, and interpret what your body has to say. Those communications will be your guide—far more than this book or I will be—to your better next steps. Still, I'll be on hand to advise and help out.

It's also why this book includes a partner—a set of resources and access to live human coaches to help you continue the personalizing process and get real feedback on your choices. The QR code for that access is included in the "Operator's Manual."

Feeling overwhelmed? I can simplify.

We all start with the same plan.

That plan's parts include guidance for what to consider in making your choices going forward. This guidance enables you to make informed choices—that's your main job here—in experimenting to see how your body responds. "What works"—what your body responds to well—you'll keep doing, and you'll glean insights from all resulting data (the experiments have no "failures") and use them to refine your subsequent choices.

I hope this sounds doable. Doable is mission critical. It's must-have.

Okay. Good. Let's go for some pizza—in this case, not actual pizza (but it could be part of your personal plan if it feels right to you). This is how I worked up plans with Jan, Stacy, Don, and, of course, Mr. Pizza Pop—you'll recall that in his case, the pizza was literal—to optimize Switch function, resulting in weight-health wins.

Your plan will be *yours*. That said, each plan starts the same way, like pizza:

Crust, sauce, cheese, and a selection of toppings.

Here are the must-haves of your Switch Optimization plan:

Foundational systems—digestion, hydration: That's the **crust**.

Nutrition: That's the **sauce**.

Lifestyle: That's the **cheese**.

Support: The **toppings** are everything else that assists.

Specialty tools are optional; those will come later.

CHAPTER 7

A TALE OF TWO PIZZAS

PIZZA #1: SARAH

Sarah made an appointment after listening to a podcast in which I spoke about personalization, better nutrition, and the four pillars—balance, quantity, timing, quality. Sarah used to have two to four lunches a day. Not that she saw it that way. She told me she often didn't have time for *any* lunch, especially on workdays. She would "just grab a bar," then have another coffee with a muffin or cookie, and later grab a snack during catered meetings or in the break room. She would arrive home famished and have another snack, then eat dinner, usually "eating way too much at night, which I know is so bad for my belly." Sarah's goal was to get rid of her belly, which had been growing steadily. A childhood athlete, she'd always been on the thinner side, and she was still in "good shape," so her doctor, friends, and family didn't pay her concerns much attention. But she was starting to struggle with health issues and battling a wardrobe that no longer fit. During that podcast,

I'd flagged two key signs of suboptimal Switch function: increasing belly weight and energy changes.

Here's the plan Sarah and I developed and how we put it into play.

Crust

Digestion and hydration. Sarah hadn't checked any of the boxes for digestive complaints, instead noting "regular, poop on waking." The aha moment occurred when she checked the box for "relying on caffeine for a morning bowel movement." We added nighttime magnesium to support relaxation, better sleep, and physiologic recovery—but also to aid in morning regularity, caffeine or not. In the morning, we had her do some stretches and drink water before she had her coffee. Sarah's hydration experiment exposed her as a hose—and she now understood the importance of becoming a sponge. Her plan included implementing hydration pit stops—8–10 ounces of water—every three hours, and once daily she added electrolytes to optimize hydration.

Sauce

We figured out that timing was the most important pillar to optimize. Nutrient balance was a close second, so we worked on them together. We stuck with Sarah's basic distribution of food across the day but organized it at regular intervals, with a focus on balance:

- We experimented with a first pit stop within two hours after breakfast and another midday. She used a delivery service as "training wheels" for three months, and from that point she made the food at home and took it to work.

- We made sure she had an alternative midday pit stop—a nutrient-balanced bar (often called a protein or energy bar)—for those days she was running to meetings or traveling, or when she forgot to grab homemade food from the fridge. This afternoon snack was a strategic choice that would pay off in the moment and later in the evening. To work, it had to score a minimum 7 on the "Better Be Delicious to Me Right Now" test (where 1 to 10 is the range, and a 10 is the most delicious bite or sip she can recall) *and* deliver nutrient needs to satisfy and reduce later cravings.

- She aimed to have her last pit stop of the day one hour earlier than when she'd been eating at night. When she had plans to go out for dinner, we worked up an at-home or before-leaving-work pit stop to support better choices during her evening meal.

Note: Caffeine was Sarah's crutch when underfueling or going without fuel altogether, and a cutback was in order. Until the results from her timing experiments were in, we deferred making a major shift to her caffeine intake. Her preliminary plan was for *now*, and a caffeine rethink was for *later*.

In sum, her sauce had three components: the pit-stop schedule with nutrient balance, the "delicious to me right now" midafternoon pit stop, and the magnesium supplement.

Cheese

Here are the lifestyle pit stops that brought it all together. Every three-ish hours, Sarah would take a three-to-five-minute movement break, such as stretching or a few rounds of deep breathing. Sarah needed a break from her work, her mind needed a break from work-think, and her vagus nerve needed her to experience some rest so it could be less overwhelmed to better fulfill its role in her ecosystem. Also at regular intervals, we scheduled pit stops to run stress checks and read her body's signals. Topping the stress list: stress about things over which she had little or no control. When her stress was elevated, Sarah experimented with ways to reduce and prevent it.

Results

After thirty days, Sarah's waist-circumference data confirmed that she had lost two inches around her midsection. She reported having more energy, sleeping better, "and people have told me I'm glowing." After three months, Sarah was fitting back into her favorite pants, and she had signed up for a triathlon and started training.

Success! But before you scan or photocopy Sarah's plan . . . a cautionary tale.

It looks like I met Sarah, figured out what she needed, gave her a set of recommendations, and she did them, and she not only smashed her goals but also changed her life for the better. A bit too Hallmark movie, even for me, a self-professed binge-watcher of those hyperpredictable movies. There are things that typically doesn't make the special and never makes the reality shows, either:

- Sarah and her story are real, but they're not "what will work."
- They're what worked *for Sarah*, at a certain point in her life.
- The story was not all one smooth, frictionless curve. There were some points where Sarah "went rogue," which I call

varying the experiment or shifting the protocol. What was key was the *overall* direction of the curve.

Here's what you can take away: The plan building Sarah and I did is what you'll do for yourself, informed by the same guidance for what belongs in your crust and sauce, what kind of cheese, and how to select your toppings, such as magnesium or a Switch activator supplement. Equally important, you will get the instructions for doing the "Better Be Delicious to Me Right Now" test, which can prove a game changer in helping you make the better choices you identify more often—as it did for Sarah.

PIZZA #2: JP

JP was referred to me by his doctor. Life was busy, and he liked the way it was going—except for his recent elevated labs and suggested medications. At forty-eight, he thought his weight was higher than it should be, and his belly certainly was bigger compared to his more athletic appearance in college. Work travel brought socializing and drinking, and at home his family and community commitments did more of the same. "Now my doctor wants to put me on meds for blood sugar and cholesterol," JP told me. "Which is scary. Time to get serious. But I'm not willing to go vegan. Or stop drinking!"

So, we got cookin'.

Crust

Although JP relied on antacids and, when he traveled, a supplement to poop, he thought of his digestion as "normal." I helped him see it was anything but, and the signals his digestion was sending were directly linked to his optimal health goals.

- We initiated a digestive tune-up focused on optimizing motility (especially during travel) and correcting the antacid dependence, the cause being reflux. We stepped down his intake of carbonated drinks and use of sugar-free chewing gum, especially during travel and in the evenings. We added a combination protocol I have established, and you will get on page 112, known as a digestive tune-up, which includes magnesium, glutamine, deglycyrrhizinated licorice, aloe, probiotics, and digestive enzymes (at dinner).
- Hydration also needed attention—his hydration experiment landed him in the hose camp, not on Team Sponge. During the day he drank coffee almost exclusively, until late afternoon when he'd think to hydrate and pound 16 ounces of water. He was getting up to pee several times during the night. We added an 8–10-ounce hydration pit stop every three hours, and an electrolyte supplement in his water just before dinner. Evenings when he had alcohol, we shifted the electrolyte water back to just before bed.

Sauce

JP's focus had been on increasing protein and avoiding carbs. However, he didn't realize that both the milk in his coffee and the beans he was eating for protein provided a fair amount of carbs, too. We adjusted his Switch Optimization nutrition plan to focus on choices that gave him a max of 30 grams of carbs at pit stops and closer to 15 grams max at dinner. We also *reduced* his protein intake at night and when he had breakfast or lunch on a plane or at meetings. (The more than 40 grams he had been eating were likely leading to three things: digestive challenges, blood sugar elevation, and fat storage.)

I proposed he try a continuous glucose monitor (CGM) for a couple weeks. The blood sugar number he'd come in with was an A1C, so I wasn't taking it as the whole story. We needed better data. I was also confident that his starting plan was going to deliver directional improvement on the blood sugar issue. His starting point was 40 percent time in range. In just fourteen days it was 85 percent (!!). Optimal for a person who does not have diabetes is 90 percent or greater. Here was hard data that gave JP—a seeing-is-believing person—the evidence to go forward knowing that his plan choices were getting a positive response from his body.

Cheese

Working and traveling a lot, JP was also sitting a lot. For his digestion, metabolism, and overall energy, he needed to move more and more often. So: activity pit stops. I told him to set a timer to get up and move around. He started out conservative, with a movement break every five hours, and a goal of every three hours. When he did work out, he focused more on cardio "for weight loss," and I conferred with his trainer, who helped him shift to strength training about three days a week.

JP's assistant got on board, working with me to implement a plan for most of the hotels where JP stayed, making available better pit stops, especially in the morning, to pivot to the gym. JP invested in an at-home scale that measures body composition, not just weight, and tracks fat and lean body mass, but he agreed to use it only once a week to avoid getting too in his head about the numbers.

Results

JP negotiated with his doctor for three months to get to improved markers without adding meds. After six weeks, he was able to report blood sugar changes and eleven pounds lost—nine from fat, just two from lean mass. He was no longer using any over-the-counter medications for digestive support, and he had no reflux symptoms. He reported that the belching, bloating, and gas he'd been treating with antacids was virtually gone, unless he "was bad."

TOPPINGS ARE PERSONAL

Finally, to make any pizza more delicious and more customized, we need toppings. The toppings *you choose*. These are the tools beyond nutrition and lifestyle choices that promote optimization in *your* body now. For Sarah, that meant mindset work. When Sarah's new choices were not delicious enough for her, she was inclined to keep looking for something else, consuming until she was beyond full. We used the "Better Be Delicious to Me Right Now" test to help her notice and appreciate the food she was choosing. Permission to go rogue was another win because for Sarah, "perfect" was at best a double-edged sword. She loved to be an A student, so she'd go for it and then rebel because she felt bored and worn out. A recovering perfectionist myself, I can spot one a mile away, but freeing others of the addiction is not my area of expertise, so she worked with a coach to help her with the mindset shifts she needed for success on her terms.

As for JP, frank conversation about what he had tried in the past and his current health status led us to semaglutide as a tool for Switch Optimization.

Additional toppings came in to address his work and travel, because as he said, "Checking quality is not feasible on a plane. And not always where I'm going, either." Environmental toxin exposure was a factor, too. Living in Los Angeles, he had had to evacuate repeatedly due to fires. We looked at detoxification and weight health from two perspectives: (1) shrinking fat cells releases stored toxins that need to be eliminated, and (2) his current toxin load was putting a strain on ecosystem operations. We worked to reduce toxin load and added support for detoxification with the supplement glucoraphanin. The big aha moment for JP was alcohol. He didn't think of it as a toxin. ("I thought red wine is good for us. Should I just drink tequila?") He got the difference between things that accompany alcohol in some forms, like antioxidants, and alcohol itself (marketing aside, it is indeed toxic to our bodies). I told him that engaging a plan to get healthy while drinking at his habitual frequency was like mopping a floor in muddy boots. He decided on drinking one night during the week, one night on the weekend, with a max of two drinks on those nights. This amounted to cutting back his alcohol intake by half, or more. The results were going to be more than worth it. In view of low HDL, elevated hsCRP, and elevated ApoB, we supplemented with niacin, pro-resolving mediators, and daily ground flaxseed or psyllium fiber.

Finally, pre-workout, we added creatine with taurine. Building muscle is difficult, especially as we age. It doesn't just require "more protein" but rather key amino acids at optimal levels. Since JP was now on semaglutide to promote fat loss, and he was increasing his strength training efforts, I wanted to double down on the amino acids that would promote muscle building and retention. That they are also a win for brain health and promoting a healthy inflammatory response was just gravy. To date, JP has maintained a loss of twenty pounds of fat mass and gained five pounds of muscle. He has not needed to add any additional medications.

The key for all my patients is the permission and the actual requirement to go out and experiment with choices—"me-search."

Me-search can include "going rogue" and having the Dodger dog, then seeing what happens with your stomach; or trialing four slices of pizza while wearing a CGM; or skipping breakfast and then trying to be nice to your colleagues or students midday. (Yes, some experiments require a warning label regarding how they may impact others.) You get the gist. Don't just believe me—or any expert saying, "I don't think that will make you feel better" or "Eat this to satisfy you." Me-search your choices. Sarah learned that late-night eating made her feel awful because when she did it after not doing so for a few weeks, the results were memorable. She changed what she got at the coffee shop after doing her stress check-ins and seeing that her afternoon caffeine made her more stressed, not less.

CHAPTER 8

THE FIRST THIRTY DAYS

It's time to make your pizza. What will make it *your* pizza is the (a) recognition that you are already doing some of the plan's choices—good, keep those; (b) you'll make new choices to align more closely with what is suggested. This combination, and tracking how your body responds to it, will give you your plan for optimal health.

As you get into the plan, you'll want to familiarize yourself with the "Operator's Manual." At points in the process of personalization, you will need to review this content to get the options for your experiments and learn how to conduct them. The list of offerings and instructions help you get a better take on what to do rather than an oversimplified recommendation like "Get more X."

When the consideration that leads to a callout applies to you and your body as of now—or if you believe it does—hop to the options in the manual to inform your choice in the plan. The manual is like having a coach—you have access to one of those too, online—who makes suggestions for you to consider as you experiment with a new or adjusted choice.

For example, the plan includes fiber minimums per pit stop. Fiber is key to an ecosystem's success, and certain fibers optimize Switch function directly. However, if you currently experience constipation, bloating, or other bowel motility challenges, or if you're on the Shot or other medications, you will want to personalize the timing and quantity of your fiber, modifying the plan's suggestion. In the digestive discussions, you receive more guidance on how to do that. The online resources that accompany the book deliver additional opportunity for personalization.

Three questions fall under the most frequently asked. We'll hit them here, to help you get going with your plan and manage expectations.

1. **There's an order to the plan. Do I follow it?**

Yes—to start. The crust has to come first. That's any pizza's foundation. Beyond that, you'll build the pizza your way, using informed choice. If you find that in the sauce or cheese your choices are already better—they align with the plan suggestions—and nothing suggests you should adjust them, then you want to keep those "a" choices as described on the previous page.

With crust in place, sauce or cheese? You can work on them together. Or if you're doing everything under one already, then you want to put your focus on the other. When to go to toppings? If you came to the plan already taking medications or supplements, or doing certain activities, you want to review those. Or if as you optimize your crust, the sauce or cheese just doesn't feel doable without toppings, you'll go there at the same time.

2. **Someone told me not to do X. What should I do?**

Who is that somebody? If something in the plan doesn't align with or contradicts your practitioner's recommendation, *don't do it.* I'm not your doctor—I'm no one's doctor—and I'm also not *your* dietitian. From my experience as a dietitian, I can give ideas for consideration—experiments to assist your body in telling and showing you what it needs (or doesn't need) and to help you gather insights to refine your plan. One of my regular recommendations is that my patients discuss with their practitioners experiments they're thinking of doing—and discuss potential results to expand the collaboration. If an item in the plan conflicts with what you've been taught by or heard from someone you know or follow, set their advice or teaching aside for now. Something can work brilliantly for one person and not work at all for others. There could be real validity to what we get from such sources. But each item in the plan is meant to become personalized to you, and an idea or suggestion made by someone else and intended for just anyone is not. Even identical twins don't get the same results from the same nutrition and lifestyle choices. Chances are the person providing input that seems to conflict with a plan suggestion is not your identical twin but rather someone who knows you socially or maybe not at all. They're talking about some pie-in-the-sky pizza that's right for them, *not* you.

3. **Do I have to read the whole plan first?**

Hello, soulmate, great to meet you. I am the one who infamously didn't read the instructions to assemble my first bike—didn't even take them out of the box. Hours later, I had to beg an instruction-reading family member to help me because it turns out a bike doesn't work without a chain. Don't be an Ashley here. Be better. Read the whole plan, then come back to build your pizza using informed choices. After all, you aren't just making any pizza—you are making yours.

ROLL OUT THAT CRUST

For your ecosystem to be optimally resourced, your digestion and hydration *have to be* functioning optimally. Here's where I contradict myself. I know I just said read everything first. But you can do this easy experiment now or first thing tomorrow to get a sense of how a major part of your Switch function and weight-health ecosystem is operating.

HOSE OR SPONGE? THE HYDRATION EXPERIMENT

Find a time when you're relatively clear of obligations and distractions for three to four hours. You don't need to be completely unoccupied, just able to track some things. Make it a time at least an hour after caffeine, alcohol, food, or intense activity and when you won't have or do those for at least another two hours. For most of us, a good time to do this experiment is on waking and after you pee and/or move your bowels. (You can take any medications. If you take medication with water, make the water you drink to take the medication the same 8 ounces you use in the experiment.)

Measure out 8 ounces (1 cup), not more, of water.

Make it the kind you usually drink—tap, filtered, bubbly, bottled, spring.

THE FIRST THIRTY DAYS

Drink it.

Note the time when you drink it.

Don't eat or drink or work out, but anything else is fine.

When you first notice the sensation telling you that you could pee, note the time.

Pee, and then go ahead with whatever you put on hold earlier—eating, drinking, exercising, and so forth.

Results: Was the time between drinking water and when you first felt the sensation that you could pee closest to:

120 minutes after drinking water, or sooner? You are a hose.

120–180 minutes after? You're a sponge.

More than 180 minutes? A hose, too.

Factors such as pregnancy, periods, medication, and others will impact but *not invalidate* the results. They may require personalization to your hydration—becoming a sponge—plan. For my hoses, the plan suggestions are designed to help you optimize and maintain hydration. If you're a sponge today, great. You may want to do the experiment at other times of the day or on different days of the week to see if there are times when your hydration needs optimization. In fact, whatever your results, repeating this experiment is a good idea. Hydration status, like blood sugar, changes with a variety of factors.

Be sure to note your result and keep it. It will resurface in the assessment and is also something you want to track for future comparison.

How Can I Optimize Hydration?

Yes, you need to drink water. But wait . . . what counts as water?

Water. It's just hydrogen and oxygen, better known as H_2O. That's the list.

For the sake of clarity, here are all the sources we've hoped, wanted, and been told count but for hydration purposes do not: popsicles; juices; "hydration" drinks; "energy" drinks; milks (dairy and plant-based); sodas (diet and regular); drinks with "water" in their names—tonic, coconut, cactus, watermelon; coffee and tea (caffeinated or not); coffee and tea drinks; foods with high water content; and anything with sweeteners (caloric or nonnutritive); or alcohol (not even kombucha). For hydration purposes, none of those is H_2O, though they should have some water in them.

The yes list is short: water—bubbly, flat, tap, filtered, spring, mineral.

For the also-rans, and to satisfy your raised eyebrows or win a dinner-table hydration bet, keep reading.

The yin to water's yang are electrolyte minerals. Just drinking water doesn't necessarily make us hydrated. There's more to the ecosystem story. Five minerals—magnesium, calcium, sodium, chloride, and potassium—combine to support hydration in the body. Their collaboration is a story of balance. Their collaborative efforts ensure that water is where it needs to be and doing what's required for optimal ecosystem function:

- Outside the cells, water activates several efforts.

- Inside the cells and in the transition from outside to inside, water escorts nutrients.

The electrolyte minerals align with these roles. Sodium, chloride, and calcium focus on efforts outside, while magnesium and potassium bring and keep water operating inside. Any imbalance of the five upsets hydration equilibrium inside and outside the cells.

Water naturally contains trace amounts of these five minerals. Treatments, filtration, fortification, and pollution impact the amounts of them and the presence of harmful things we call "unwanteds."

Your hydration strategy unfolds in three parts: (1) Optimize water intake, (2) optimize electrolyte intake, and (3) adjust dehydrators.

1. OPTIMIZE WATER INTAKE

- Balance—choose water that naturally retains a better balance of electrolyte minerals or upgrade your water by adding them with a food or supplement source of these hydration-assisting nutrients.
- Quantity—as a starting point, drink 40 ounces of plain water per day. Depending on your size, health, and goals, more may be a win.
- Timing—spread your water intake over your day. Just as pit stops about every three hours serve well for nutrition, intervals of about the same length work with water. Pounding water 40 ounces at a time isn't better; neither is taking in a few ounces every ten minutes.
- Quality—choose water that is as free from unwanteds and

away from dehydrators, whenever possible—in sources and packaging as well as in the water.

2. OPTIMIZE ELECTROLYTE INTAKE

- Drink water that retains or is fortified with electrolytes.
- Spring and mineral water retain electrolytes. There is a wide range of amounts based on their sources, but if they are filtered, they will have fewer.
- Fortification with electrolytes carries a buyer-beware note, both for your wallet and because the quality of the minerals (the form) and their individual quantity can be helpful toward a balanced amount or can create or exacerbate existing imbalance. More on amounts of minerals and forms in the toppings section of the plan in Chapter 12. Consuming electrolyte minerals in excess can be harmful.
- Consume foods that deliver electrolytes in balance.
- Mother Nature tees up electrolyte balance for us by offering foods in addition to water that contain some or all five electrolyte minerals. It is very difficult to OD on these minerals from foods. A more common issue is that your food provides an insufficient amount.
- Similar to above, fortification and food processing significantly alter the amounts and forms of electrolyte minerals. These foods, like the supplements below, can absolutely contribute to producing a conditional insufficiency or frank deficiency of one or more electrolytes.
- When supplementing in general, it is vital to consider total nutrition—what your food delivers or doesn't—and any

existing nutrient gaps. We will go over this in the toppings chapter in detail, but this is why the one-size-fits-no-one hydration packets create problems. Not everyone needs to add more sodium along with magnesium, chloride, and potassium to improve hydration. Many need to increase their intake of all the other minerals and may need to reduce their added sodium from dietary choices, too.

- Choosing better electrolyte supplements includes evaluating the active nutrients for forms and amounts that are better for your body and in forms the body recognizes and that do not include "other ingredients" that challenge the ecosystem.

- Whereas it is difficult to give the body too much of one mineral or even throw off the balance of your electrolytes with food (not including added salt), it is very easy to do so with supplements.

- Evaluate your sodium intake before you add an electrolyte powder that includes sodium—especially many of the ones available today that have 1,000 milligrams (1 gram) per serving. Between packaged foods, sauces, prepared foods, beverages, and restaurant meals, as well as other supplements (protein powders, etc.), check to make sure you aren't already getting in more than 3 grams of sodium daily.

- Likewise, proceed with caution in adding more chloride, which is found in a lot of foods, often along with sodium.

- Evaluate your calcium intake to make sure you aren't exceeding 1,000–1,200 milligrams (1–1.2 grams) based on age and health, as excess will make it hard to get the optimal impact from your current magnesium intake, an imbalance of which will be a factor not just in suboptimal hydration but also in other weight-health issues.

- Supplements can alter one or more electrolyte levels, creating a need to adjust overall intake for better balance and hydration.
- Medications and treatments can impair absorption of electrolyte minerals, increase needs for electrolyte minerals, or make an increased intake of one or more electrolyte minerals problematic.
- Address any medications, supplements, activities, or health issues that disturb electrolyte balance.
- Acute or chronic illness, when the body is challenged, impacts hydration—sometimes severely (fever, diarrhea, vomiting, etc.) and other times more moderately but consistently (headaches, pain, etc.).
- Hormonal shifts also impact hydration via changes in thirst, urination, sweating, and stress.
- Activities—whether and how much we sweat during them— will impact our hydration needs.

3. ADJUST DEHYDRATORS

Get ready! This next list may read like "I knew it; she's going to make me ditch all my favorites." From my patients' stories— and my own—you know that's not how I roll. But I also know that these do impact ecosystem and Switch function. What you will now learn is how they impact your hydration. These insights will help you optimize your hydration. The goal is to adjust the timing, amount, and quality, as necessary, of those that impede your ability to become and stay a sponge.

- Caffeine—in coffee, tea, herbs, soda, and energy drinks—is a diuretic, which challenges hydration. Impact depends on the source, your genetics, and other dehydrators. The key takeaway is that caffeine drinks, even if made with water, are not part of your hydration strategy. Separate your caffeine intake from your efforts to optimize hydration (as explained above) with water and electrolytes.

- Sugar in the bloodstream triggers ecosystem activity that results in cellular dehydration. The goal is to make sure that not only sugar but whatever else is being transported in the blood (including nutrients and Switch hormones) is able to move through without getting stuck. When a higher concentration of sugar in the bloodstream is detected, the cells release water into the bloodstream. This upsets your cells' hydration status. We see this with high blood sugar, where dehydration is a side effect.

- There are several ways alcohol impacts hydration. Let's focus on its inhibition of the antidiuretic hormone (ADH), also known as vasopressin. Your kidneys do not get the message to retain water, so you pee. This results in more frequent urination and loss of water and electrolytes—ecosystem disrupted. Your Switch Optimization plan may include alcohol, though I suggest for the first thirty days you abstain, especially if you are working to become a sponge. Another reason? You are about to learn how alcohol disrupts your digestion, specifically the lining of the digestive tract, the residence of your Switch hormones. My experience with my patients reveals four groups—those for whom alcohol was never a factor, those who eliminate it for good, those who significantly reduce it (maybe do a trial elimination to kickstart), and those who elect to remain consistent alcohol drinkers. Of the latter three groups, those who eliminate it and those who significantly reduce it experience more efficient and effective weight-health-ecosystem optimization. It's worth noting that the general recommendation when on the Shot is to eliminate alcohol. In the coming sections, and online, you get support and strategies for alcohol elimination.

- During periods of stress such as high-intensity (for you) exercise, being stuck in traffic, or amid a significant argument, the body turns on its stress response. The stress response uses water and electrolytes, so the experience of stress increases fluid and electrolyte use. This results in depletion—or dehydration. In the coming sections, you will get strategies to assess and optimize your stress response. What you will not get is the "you need to reduce your stress to be weight healthy," one-size-fits-no-one standard recommendation.

WHAT IS YOUR DIGESTION SIGNALING?

It is easy to not notice if your hydration is off, but not your digestion.

Our digestive system constantly sends us signals and will increase their intensity and frequency until it gets our attention. So if I asked you, "Is your digestive system sending you signals that it's functioning suboptimally?" your answer may be a straight yes. For most people, it is. And yet we've become very accustomed to ignoring, overriding, finding work-arounds for—even normalizing—suboptimal digestive function.

Digestive function is at the core of ecosystem operations and, as we've discussed, where your Switch hormones reside. Your digestion is as personal as personalization can get. As you've learned from my story and my patients' stories, if your body can't use what it is getting, then your weight-health outcomes will be challenged. This makes digestive optimization an essential in the ongoing

GPS ALERT

ONE FOR THE GUYS

Since many women experience digestive disturbances monthly for several decades of their lives (including during pregnancy), and because societies have historically been more critical of women's waistlines and bellies, women are more inclined to note and report digestive symptoms and attempt to treat them (not necessarily better) than men.

So, guys, I'm not trying to pick on you, but I am pretty sure that many of you, like many of my patients, write off suboptimal digestive function as "normal" and don't register it as what it is: a key ecosystem performance indicator. Those who feel that *in*digestion is no big deal, gas is just gas, bloat is just a "life after thirty" thing, and reflux is earned ("Should've stopped eating and drinking sooner") are ignoring the earliest signals of conditions and diseases that blindside them. One of these conditions is the weight-health issues that brought you here. Digestive issues are your ecosystem and your Switch signaling that there's something wrong under the hood.

effort to become and remain weight healthy. All phases moving forward—assessment, experimentation, optimization—include discussion of and strategies for digestive health.

Sometimes the digestive signal is a bit hidden. Recall Sarah. At first she felt pretty good about her morning poop schedule. Then, as she checked the box noting that she relied on coffee for that first bowel movement, we got an insight about her digestion. Just because an action is happening doesn't mean it is happening optimally. Let's decode your digestive signals using the table on the next page.

If you answered yes to any of those questions, addressing your suboptimal digestive function is a foundational part of your Switch Optimization plan. It will be baked into your crust.

IS MY DIGESTION FUNCTIONING OPTIMALLY?

Go with what better describes you and your body today or most days over the past week.

	YES, I HAVE	NO, I DON'T HAVE	I DON'T KNOW
Gas, bloating, burping, reflux			
Constipation, difficulty moving bowels, hemorrhoids, poop looks like pellets or is very hard			
Rely on a "helper" to poop (smoking, caffeine, supplements, herbs/herbal tea, etc.)			
Loose stools, diarrhea, uncontrolled bowel movements			
Food allergies, intolerances, ingredients/foods/food groups avoidance because they challenge my digestion			
Acne, skin irritation, rosacea, hives			
Body and mouth odor			
Address digestive concerns by taking supplements			
Take prescribed or over-the-counter medications to improve my digestion			
Current diagnosis of digestive and/ or autoimmune disease, conditions, syndromes			

THE FIRST THIRTY DAYS   113

As you proceed, note that the remainder of the plan suggestions—sauce, cheese, and toppings—also support digestive optimization.

ON TO THE SAUCE

The thing about sauce—it's gotta be delicious. That's number one, so we will start there. But the sauce plays many roles in your better pizza, so we will also dive into those. They include the sauce acting as a protector, balancer, and nutrient-value addition.

But the taste of the sauce will make or break a pizza—literally or, as we are discussing here, figuratively—so let's start there first. Often—make that too often—our nutrition choices are swayed by factors such as convenience, people-pleasing, and what we perceive to be or is advertised as "healthy." And we compromise on taste. Turns out, that's a problem for our Switch hormones and ecosystem. Even before the nutrients we consume are audited by the digestive system, a phase occurs where multisensory data triggers the early release of some gut hormones. This starts a gut-brain conversation focused on intake of nutrients for use and becoming optimally resourced, which the gut-brain equates with satisfaction.

Some GLP-1 is made in our taste buds, where there are conveniently receptors for it, too. Thus, the taste buds sense and send anticipatory "food receipt" messages when fat and sugar are consumed. When we taste things that are delicious to us, our brains get "I am satisfied" directly from our mouths.

In those with suboptimal Switch function, an apple's sweetness may be perceived as not sweet enough to trigger satisfaction—instead triggering cravings for something sweeter. This helps to explain why intake of nonnutritive sweeteners—delivering exponentially higher sweetness—impacts Switch function, driving the demand for higher levels of sweetness for satisfaction. Both of these sweet scenarios highlight why oral health—part of digestive health—is also a key Switch performance indicator.

What does this mean for you and your choices?

Satisfaction is more than the feeling of fullness.

Your brain will get those satisfaction messages more effectively and efficiently from what it senses—sights, smells, and actual tastes—as delicious to you *in the moment*. When what we consume is not delicious, we tend to overconsume our current choices, hoping that quantity sums up to a satisfied state; or we go looking for something tastier. In the decade of the "I will have a salad with grilled chicken, no croutons, and dressing on the side" lunches, my patients consistently reported intense cravings for desserts, snacks, and sodas throughout the afternoon and evening following that less-than-delicious-to-them meal. This frequently reported dissatisfaction experience led me to devise a game that now is a staple tool of my program.

THE "BETTER BE DELICIOUS TO ME RIGHT NOW" TEST

Start with a clean mouth (not post-coffee, but also not right after brushing your teeth or using mouthwash; you can have a sip of water or a bite of a neutral-tasting food, such as a cucumber, to render your mouth "clean").

Devise your scale for assessment—from 1 to 10 where 1 is "*Ick*, hold my nose, don't want to have that *ever* again," and 10 is something you could sing or write a sonnet about. Try to get as specific as possible regarding your choice, recalling its smell, appearance, and taste.

Choose a food or beverage that you plan to consume in the coming moments.

Take two bites or sips of that choice.

Pause and rate. What score do you give your bites or sips?

If it is a 7 or greater, carry on and enjoy.

If it is a 4, 5, or 6, think about whether you have other options. Are any of them better—strictly by the standard of deliciousness—than this one? *Note:* Coming up, you will get guidelines for how to compare choices by different standards.

If it is a 3 or less, put it down and see whether there's something better. There may not be, and that may be your cue to stop and ask, "Do I need fuel at this moment?" This helps you distinguish appetite from hunger. If it's appetite, as in "No, my body doesn't need fuel right now," move on, choosing not to have something less than delicious to you right now. If it is hunger, you should consume something. Your deliciousness results are a flag to let you know that your fuel choice is unlikely to fully satisfy you. You may need to use other guardrails to ensure you don't overconsume or start a search for something more satisfying before your body needs refueling. An additional strategy to avoid overconsuming and reduce future temptations is to place a specific portion on your plate, and once consumed, clear your plate; upon finishing, get up from the table, leave the kitchen, or do not walk by the break room or a desk with a candy jar. Aim to have your next pit stop score better.

This test's a game changer, especially when you're eating out—restaurants, parties, travel—and in any circumstances that limit your control over the quality of your choices. When there are fewer "delicious to me right now" choices, a frequent result is overeating or overdrinking.

We are here to make delicious sauce. But our sauce, too, has additional goals:

To be a vehicle for optimally resourcing digestion and hydration

To optimally resource the Switch and ecosystem via the four pillars of better nutrition

A conventional approach would be to provide you with two lists:

Eat this—not that.

Eat now—not then.

Optimization is unconventional. It works like this:

START YOUR CALORIC WINDOW:

- Start your calories by *no later than* 10 a.m. and finish by 8 p.m.*
- Trial limiting your calories (including caloric-containing supplements) to a ten-to-twelve-hour window (like 8 a.m.–6 p.m., or 7 a.m.–7 p.m.).**

*Adjustable based on your work/life schedule; the goal is to have calories within three hours of waking and stop three hours before going to bed.

**Unless you have other practitioner recommendations or personal health insights.

DURING YOUR CALORIC WINDOW:

- Plan a pit stop for about every three hours.
- Per-pit-stop (meals and snacks) guidelines:*
  - Carbohydrates: 15–30 grams
  - Protein: 15–30 grams**
  - Fats: 5–15 grams**
  - Fiber: more than 7 grams; aim for a variety of fibers
  - Water: 8–12 ounces

- ◆ Unlimited non-starchy vegetables, herbs, spices
- ◆ Sweeteners: Less than 2 teaspoons daily of sweeteners (1 teaspoon is about 4 grams of sugar); avoid all artificial nonnutritive sweeteners and keep "natural" nonnutritive sweeteners to one serving daily (includes stevia, monk fruit, allulose, erythritol, xylitol, etc.).

*Applies to your food + supplements + beverages.

** If you go higher, ensure digestion is optimal to break it down and use it.

- Choose a variety of carbohydrates, protein, fats, and non-starchy vegetables to support greater nutrient diversity.
- Aim for a rainbow of colors from herbs, spices, fruits, and vegetables daily.
- Include at least one serving of Switch-optimizing nutrients daily:
  - ◆ Cruciferous vegetables
  - ◆ Citrus
  - ◆ Resistant starches
  - ◆ Cacao/cocoa, coffee, green tea
  - ◆ Blueberries, pomegranate, beets
  - ◆ Extra-virgin olive oil
  - ◆ Turmeric
- Reduce or avoid food or beverage ingredients that may challenge one or more components of the ecosystem.
- Aim to eat less gluten-containing grains (wheat, barley, rye, spelt) and refined white flour products (rice, rice puffs/cakes, tortillas, crackers, puffed cereals, white bread, etc.). Choose organic, whole grain (like whole wheat flour for pasta, crackers, bread), and sourdough varieties when possible. Oats are often cross-contaminated with gluten, so look for gluten-free oats if possible.
- Prioritize eating fruits (and vegetables) with their skin when possible (for example, apples, pears, cucumbers, and potatoes).

In sum, practice each of these choices with the "Better Be Delicious to Me Right Now" test a few times a day so that it becomes a regular strategy.

CIAO, PYRAMIDS AND PLATES; HELLO, PILLARS

We've unlearned the purported values of diagrams—pyramids and plates—in favor of plans that come together via personalization. In the patient stories thus far, you've seen how we personalize a plan: pizza consumed in smaller amounts, more often; snacks or lunch are both pit stops that we optimize; alcohol doesn't have to be zero, but better when reduced in amount and frequency; and so forth.

How do we personalize your total nutrition plan? We start with figuring out what's better—and not—in your current nutrition choices. To do that, we need to conduct an assessment—compare your current intake against the four pillars.

THE FOUR PILLARS ASSESSMENT

You will want to reference the "Switch Optimization Nutrition Plan Recommendations" in the "Operator's Manual" to help you complete this assessment.

Pick a day or use yesterday and note what you ate, drank, and took (supplements), and when.

Note the date, time, and your answers when you are collecting data so that you have your answers for easy reference when you reassess.

Assess your choices against the four pillars of better nutrition, which here we will look at in order of quantity, balance, timing, and quality.

Quantity

Here we assess how much of a macronutrient or other food and nonnutritive ingredients you are giving your body at one time.

Looking at each pit stop independently, determine if your total consumption (food and beverages) was inside or outside these ranges (remember, better is not perfect, but 35 is closer to 30 than 40):

- 15–30 grams of carbohydrate
- 15–30 grams of protein
- 5–15 grams of fat

Answer yes if any of the following are also applicable:

- I ate fewer than two colors of the rainbow (from fruits, vegetables, herbs, or spices, not added colors).
- My total added sugar from foods, supplements, and beverages was more than 2 teaspoons (1 teaspoon is 4 grams, so 2 teaspoons is 8 grams of sugar).
- My total "natural" nonnutritive sweetener intake from foods, supplements, and beverages was more than one serving (1 packet, 1 spoonful, 2–3 drops, or 1 serving size per label of stevia, monk fruit (Luo Han Guo), allulose, erythritol, sorbitol, maltitol, mannitol, isomalt, lactitol, etc.).
- I consumed a serving or more of alcohol.

Balance

Specifically, here we are looking at the combination of your macronutrient—carbs, fats, proteins—intake to see if you balance them each time you fuel up.

Looking at each pit stop independently, was your total consumption (food, beverages, and supplements) at each pit stop not in balance?

Answer yes if:

- Your number of servings for each macronutrient was not the same (example: 1 carb, 4 protein, 3 fat).

Answer no if:

- You had the same number of servings for each macronutrient (so in total at your pit stop: 1 carb, 1 protein, 1 fat, or 3 carbs, 3 proteins, 3 fat servings).

Timing

Your caloric window is the period in the day over which you take in calories via both food, beverages, and supplements. Answer yes if any of the following apply to you most of the time:

- On waking, I don't feel hungry for at least one hour.
- I typically wait for more than three hours after waking to consume nutrients.
- I drink noncaloric beverages (coffee, tea, flavored water) to keep from needing breakfast.

- Before I have breakfast, I have to ignore or distract myself from hunger.
- When I do have breakfast, I overeat because I am really hungry.
- Following breakfast, I often go less than three hours or more than three hours without eating or drinking (a caloric beverage).
- I eat or drink caloric beverages less than two hours before I go to bed.
- Most days my caloric window is less than ten hours or greater than fourteen hours.

Now, evaluate your answers on quantity, balance, and timing to identify your better next steps.

For any question you answered with yes, consider that pillar one to optimize in the coming weeks. It gives you insight on where to focus. For questions that you answered with no, you appear to be in good shape with respect to that aspect of that pillar. Take a moment and review your answers for these three pillars and make notes toward your action plan, considering choices you think will help in meeting your body's current needs.

Quality

Quality, the fourth pillar, we're going to look at differently. There are two categories of quality:

- Quality we know
- Quality we don't know

We can assess the quality we know: Is it better or not? The or-nots, especially those we consume regularly, we identify for upgrade.

We can't easily assess the quality we don't know—coffee and a sandwich at a coffee shop or a meeting; food you're served in a restaurant.

When developing your plan, we'll assess the quality where feasible. Where not feasible, we'll assess the share of your intake that comes in as quality unknown. Right now, let's familiarize ourselves with how to recognize better and not-better quality choices.

Think about the things you consume most often—foods, beverages, supplements, and their ingredients. Familiarize yourself with the table on the next page. The left side is the list of items that from a quality standpoint are likely a better choice. The hit list, on the right, covers stuff to reduce where possible. My mantra, with all things in life but especially nutrition, is "Better, not perfect." The goal isn't to avoid all sources of not-better quality but to be aware of them and work to reduce them, especially in what you consume more often. Consider how often your choices include ingredients from the "Not Better" column. In the next section, you will get guidance on replacements.

OPTIMIZING YOUR NUTRITION— THE FOUR PILLARS

Balance

Homeostasis—you've probably heard the term or been quizzed on it at some point. It's the body's big, audacious goal. But because the body is a dynamic being, it never quite achieves the balance it craves and was designed for. By design, the body wants balance in

GUIDANCE FOR FOOD
AND SUPPLEMENT QUALITY

When it comes to optimal health, your body will run better when it gets what it can recognize and easily use more often. This is a tool to help you identify better and not-better choices in foods, beverages, and supplements. Use it to identify what may be impacting your body's operations and what upgrades may help.

Quality Standard	Better ✓	Not Better* ✗
3rd Party Testing	USDA Organic, Non-GMO Project Verified, NSF, USP, cGMP, Glyphosate Residue-Free (Detox Project), Certified Gluten-Free, IFOS (fish oil only), Fair Trade Certified, Demeter Biodynamic, Animal Welfare Approved; Marine Stewardship Council (MSC); Best Aquaculture Practices (BAP); EWG Food Scores; companies making Certificate Of Analysis (COA) available	No 3rd party testing; branded (their own) test results exclusively; marketing terms (organic fish, GMO free, etc.)
Meats, Poultry, Cheeses	Humanely raised, without hormones and antibiotics, grass-fed/finished; meet the other quality standards	Blackened, grilled (black); highly processed meats (cold cuts, hot dogs, etc.) and cheeses**; don't meet other quality standards
Fish	≤0.1 ppm mercury; <10mcg mercury/3oz serving; selenium:mercury >1; best aquaculture practices; pregnancy safe; no detectable microplastics; meet other quality standards	Not 3rd party tested or exceeds levels; don't meet other quality standards
Fruits, Vegetables, Herbs, Spices	Frozen, ready-to-eat, dehydrated, canned (in water, olive oil, rinsed), meet other quality standards	Don't meet these or other quality standards; not 3rd party tested / no available information
Dairy, Non-Dairy, Nuts, Seeds & Oils	Whole, full fat, low fat, fermented / cultured, extra virgin; cold-pressed unrefined /mechanically refined nuts & seeds (raw, roasted, toasted), whole, pieces, as butters (other ingredients meet quality standards)	Hydrogenated, partially hydrogenated, modified fats and fat-free; "vegetable," corn, cottonseed, canola***, soybean***, safflower***, sunflower***; don't meet other quality standards
Grains, Flours, Legumes	Whole grains, beans, lentils, peas, peanuts	Enriched, white flours; puffed; rice cakes; pieces and parts (vs whole) like starches, brans, etc. (regular or modified food starch)
Beverages, Yogurts, Protein Powders	Unsweetened, meet other quality standards	Excess added sugar, nonnutritive sweeteners; don't meet other quality standards
Sweeteners	Whole fruit and vegetables; organic honey and sugars, syrups (coconut, maple, molasses, date) and natural nonnutritive sweeteners (stevia, monk fruit, erythritol, allulose, xylitol, etc.) from sources that meet quality standards	Artificial nonnutritive sweeteners; sweeteners that don't meet other quality standard including non-organic sugars, high fructose corn syrup, corn syrup, agave, glucose-fructose, and fructose-glucose

©The Better Nutrition Program

Quality Standard	Better ✓	Not Better* ✗
Colors, Flavors, Preservatives	Colors, flavors, preservatives from food ingredients	Artificial colors and flavors; gums and thickeners** [note: the sources of "natural flavors" are often unknown; use the better choices more often]
Processing: fortification, refined, enriched, decaffeination	Manual "can do in your kitchen, doesn't require a chemistry lab" processing that does not alter the original food form to remove foundation ingredient(s) [examples: blending that retains fibers, dehydrating a fruit etc.]; adding ingredients — like DHA, folate, lion's mane etc. —that meet quality standards; Chemical-free decaffeination; "other" and inactive ingredients meet quality standards	Chemical (chemistry lab required to make ingredients or do processing) that alters the original food form to remove foundational ingredient(s); Fortifications with ingredients that do not meet other quality standards (folic acid, carrageenan, artificial sweetener etc.); chemical decaffeination
Heavy Metals; Mycotoxins	Third party testing available; below FDA action levels; state-specific standards met such as "safe harbor" levels (CA Prop 65 which is often used as a 'better standard' but only require a warning on product in CA); USP verified	Not 3rd party tested or exceed levels
Microplastics	No detectable microplastics	Not third-party tested for microplastics; marketing claims free from or zero microplastics
Containers, Utensils, Tools, Storage	Stainless steel, glass, paper, bamboo, wood, cast iron, dark glass for ingredients that are sensitive to light; sealed until used for those sensitive to oxygen	Bins and bottles are routinely opened and closed or remain open (for nuts, seeds, grains, oil); clear containers for oils and oil-based sauces; stored in cans & packages with BPA lining; heated in plastic

*For example, food and food packaging ingredients: artificial nonnutritive sweeteners: Sucralose (Splenda®), Saccharin, (Sweet'N Low®, Sugar Twin®), Aspartame (Equal®, Nutrasweet®, Sugar Twin®), Neotame (Newtame®), Acesulfame-Potassium (Sweet One®, Sunett®), Advantame. Natural nonnutritive sweeteners: Stevia (reb A is an extract), Monk fruit (Luo Han Guo), Allulose, Erythritol, Sorbitol, Maltitol, Mannitol, Isomalt, Lactitol etc.)sPhenylalanine, caramel, synthetic dyes (blue nos. 1 and 2, green no. 3, red nos. 3 and 40, yellow no. 6), tartrazine, sodium nitrate/nitrite, BHA, BHT, potassium bromate, paraben/propyl paraben, benzene, ethylene chloride, titanium dioxide, brominated vegetable oil (BVO), BPA, phthalates, and carrageenan

** These meats and cheeses typically contain added colors, flavors, and preservatives (like those listed below) and often are processed using chemicals. Note that processes such as grilling, curing and smoking can expose you to an additional alphabet soup of chemicals – HCAs (heterocyclic amines), PAHs (polycyclic aromatic hydrocarbons), NOCs (N-nitroso- compounds)

*** If you consume choose organic and high-oleic versions, avoid regular use where possible nutrients, tastes, and textures. Your body has needs—lots of them—and they work together, not in isolation. That's why it makes zero human health sense when one nutrient or type of activity gets more attention than another. Yes, fuel goes into the tank and air goes into the tires—but a major factor in gas mileage is tire pressure, so they both need to be optimal for the car to run better. The body runs on countless synergies like this, only more complex. Your body's fuel (carbs), air in the tires (protein), and oil in the engine (fats) cooperate to help you run optimally throughout your day.

No whole food exists in nature as a single macronutrient. Not. A. One.

- Beans, grains, and some dairy have carbs and proteins.
- Eggs, meats, fish, and some dairy have proteins and fats.
- Fruits and vegetables may have carbs, proteins, and fats, for example, an apple, has some amino acids (proteins) and fatty acids (fats).
- Even butter—the ultimate fat—has amino acids (proteins).

If you identified balance as a pillar to focus on, let's review the following and see how you can experiment to improve the balance of your choices:

- If you have an 8-ounce latte with whole milk, the body is thrilled you gave it protein, fat, and carbs, but less thrilled when you throw at it more carbs—so many more—in the form of oatmeal, fruit, and maple syrup.
- If you have beans, the body is getting protein, carbs, and fiber, so it doesn't need chips, corn, rice, fruit, oat milk lattes, juice, or plantains, too. It wants you to have guacamole,

meat or eggs or fish, maybe a little cheese, some water, and loads of non-starchy vegetables, nuts and seeds, and herbs and spices.

- If you serve the body a burger—beef, bison, or salmon—it really wants some greens, mushrooms, onions, even mustard, sweet potato fries or a bun, black beans or beets; but it feels pretty overwhelmed if you serve it an egg or bacon on top of the burger for more protein (and getting more fat). If you want to round out the meal with a dessert, your body would go for more colors in the form of berries, upgraded with a scoop of ice cream, in a trade-off for the bun or fries.

Here's an analogous balance story that plays a bit differently. Two comparative days, looking at how choices balance:

NOT BALANCED	BETTER BALANCE
Oatmeal with fruit, coffee with cow/oat milk (8 ounces)	Oatmeal with nuts and seeds (nut/seed butters); option to add cacao nibs and spices; coffee with unsweetened almond/hemp milk (if you prefer the oat or cow's milk, then choose a different breakfast option, such as eggs or cottage cheese with berries)
Sandwich with turkey and cheese; sparkling water	Half a sandwich with extra turkey and avocado; cut cucumbers with yogurt dip; sparkling water

NOT BALANCED	BETTER BALANCE
Sweet tea/coffee latte, "energy" bar: 45 grams of carbs, 5 grams of protein, 7 grams of fat	Unsweetened iced tea, "better energy" bar: 15-30 grams of carbs, 15-20 grams of protein, 7-10 grams of fat
Water, salmon with potatoes or corn and sautéed spinach, ½ pint of ice cream or 1 popsicle	Water; salmon with sautéed spinach, tomatoes, and onions (if you want the potato or corn, pass on the ice cream); 1 scoop ice cream or frozen berries and 1 ounce dark chocolate

While foods have multiple macronutrients, most contain less than what we deem a full serving of each of the macronutrients, so we count and categorize them accordingly. As an example, we get protein in grains such as oatmeal and quinoa, but each suggested serving contains 15 grams of carbohydrate and only 2 to 3 grams of protein, much less than the suggested serving size minimum of 15 grams of protein. Thus, these grains get listed under "carbohydrates." To meet your protein needs in a meal or pit stop that includes them, pair a serving of oatmeal or quinoa with a serving of a protein and a fat rather than with more carbohydrates. (To see how foods are categorized, including categories with multiple macronutrients, see the nutrition plan in the "Operator's Manual.")

Looking at your pit stops, how could you balance them better more of the time?

A scene from the '90s played out in my office recently, reminding me that our fat-phobic roots run deep. I was helping a patient with perimenopause-attributed weight-health issues. Her complaints included dryness; fatigue; declining quality of hair, skin, and nails; belly weight gain; and elevated levels of unhealthy inflammation (hsCRP greater than 1.0; hers was a 5.0). I suggested we experiment with hemp seeds—"the wild salmon of the plant kingdom"—for protein, fats, fiber, and minerals; and adding a fish oil with specialized lipid pro-resolving mediators. She morphed into a twenty-year-old circa 1995 I knew all too well. "Ashley, they (hemp seeds) have a lot of fat. So many calories. It's going to make things worse, not better!" So, we had the BIG. FAT. CHAT.

"Like me, you were raised in the diet and fat-free era," I said. "Not only was it a disaster for our blood sugar but it also starved our hormones, hair, skin, nails, and gut lining. We need these fats. They will help you burn fat, and look and feel better." Today I would add that a failure to consume fats hampers Switch and ecosystem operations.

Another patient had sushi for lunch most days. He was trying to lose belly fat, so he had gotten rid of carbs and would eat sashimi, cauliflower rice, and steamed vegetables. But he soon paid for it with irresistible sweet cravings and energy crashes in the afternoon. His CGM showed his blood sugar was great through the meal but then dropped two hours after, which he would respond to with office candy or Diet Cokes (multiple). Adding some organic edamame (soybeans) to his meal, or a bowl of berries as his dessert, helped his blood sugar stay in range and kept him from needing the post-pit-stop pick-me-ups.

One of my least favorite terms has always been *portion control*. First, it conjures up those prefab meals in a container from WeightWatchers or Jenny Craig; second, and so much worse, it reinforces the myth that all that matters is the amount.

One of the superpowers of nutrient balance? It ushers in portion control 2.0. Nutrient balance encourages the body's system of satiety, so the body knows it's in full receipt of the resources it needs at that pit stop. For my "sashimi, cauliflower rice lunch" guy, it wasn't him but rather his body that was crying out for energy.

So what about keto or paleo? What about lower carb plans? What about higher protein needs? Do you personally feel the need to eat imbalanced to achieve your desired health outcomes?

Maybe.

Balance does not look the same for everyone, and not even for one person at all points in their life.

With your balance assessment completed, you now have the tools to optimize your choices for better balance.

Quantity

Do you ever wake up or get to work and find yourself immediately confronted by *way* too much to do? Panic, dismay, and fear kick in. Next comes irritation and overwhelm. Those moods and mental states are the pillars of paralysis. They impact how you perform. Tasks take longer to do and get done less well. You look for a way to cross off an item that's really not vital or push one that's non-urgent to another day. While you're working on your to-do list, *no work gets done*. Things fester. Soon enough you will consider quietly quitting. Or if you're your own boss, you'll consider cutting yourself from the payroll. For most people, overload like this is a regular occurrence. It has a severe impact on our performance. It drags down confidence, engagement, and self-esteem.

When we overload our body, it responds much the same way. It, too, winds up as a workplace casualty or rebels in other ways.

This rebellion is everything *but* quietly quitting: excess body fat, metabolic dysfunction, and digestive complaints, oh my!

Mainstream nutrition offers its one-size-fits-all solution: *portion control.* Diet programs and guidelines bully us with "Call on your innate willpower (and grow up)." That sells diet books and plans. It ignores how the body works. We aren't built with a "willpower" system. We *are* equipped with a Switch that directs a weight-health ecosystem. And that's what we're here to fix.

When the body gets what it needs:

It can deliver to meet all demands and requests.

It doesn't have to prioritize where to send resources, depriving itself over here to cover a more pressing need over there.

Systems that need operational support at the moment get what they need, and things that are nonurgent but "nice to have" for cleanup and prevention can get their needs met, too.

On the flip side, with better quantities and no excess, stuff doesn't get "set aside for a rainy day"—that is, stored as fat.

Your Quantities: How You Make Them Better

Your yes answers in the quantity assessment identify pit stops when your quantity of a nutrient under- or overdelivers. Those are prime opportunities for you to experiment by adjusting choices or amounts. To evaluate how your body feels about any new quantities, use your body's signals: mood, digestion, energy, weight composition, hair, skin, nails, and so forth. Reminder: At any pit stop, you're looking at your total nutrition—food, beverages, *and* supplements.

The following two quantity case studies may feature you if you answered yes to:

- Taking in *less* than 15 grams of protein at a pit stop. We want to up it to at least 15 grams, possibly up to 30 grams. The pit stop in question is avocado toast with an egg, which currently delivers 10 grams of protein: 2 grams from the toast, 2 grams from the avocado, 6 grams from the egg equals 10 grams.
    - ✧ Some options to get to 15–30 grams include: another egg, 6 grams; ½ cup of cottage cheese, 14 grams; ½ cup of silken tofu, 9 grams; 3 ounces of chicken sausage, 14 grams; 2 tablespoons of almond butter, 7 grams; 3 tablespoons of hemp seeds, 9 grams.
- Getting *more* than 30 grams of carbohydrates at a pit stop: We need to check out how your body responds when choices align to put carbs in the range of 15–30 grams per pit stop. You've shifted breakfast to a smoothie, to get your protein in. The base of the shake—a cup of oat milk—is 16 grams. A ½ cup of blueberries, 10 grams, keeps things aligned. The issue is the banana: 30 grams all by itself.
    - ✧ Some options: Switch to unsweetened almond milk (1 gram) or coconut milk (2 grams) and go with half a banana, and in place of the other half of the banana, add a ½ cup of frozen cauliflower (1 gram; you won't taste it as long as its frozen, but it gives a great texture and delivers as much potassium as the half banana); or drop the banana entirely and just do the ½ cup of frozen cauliflower.

A listing of a variety of foods and nutrients, with quantities noted, is available in the "Operator's Manual."

Timing

Few case stories I know illustrate timing as powerfully as Mr. Pizza Pop's, who you might remember ate an entire (actual, not metaphorical) pizza every day. Without making significant changes to the other three pillars—quantity, quality, balance—he achieved dramatic positive results mainly by shifting his nutrient intake from backloaded—all at the end of his day—to a sequence of pit stops starting about an hour after he got up and then every three hours up to a few hours before he went to bed. In twenty-plus years I have not met up with another client with exactly the same dietary profile, but his story shows us two good templates for timing: pit stops and a caloric window.

To identify and optimize your better caloric window, here are the rules:

- Calories—not caffeine or supplements—are your body's source of energy. Your body will call you out if you try to cheat it with replacements too often.
- Not taking in anything gives your digestion and your mind the rest they need. But if there is too much time without fuel, they start to freak out. There's better timing for these time-outs; discover yours.
- At different life stages and on different days within a life stage, your body may do better with a different nutrient window. **So, define a window, but don't plan to stay with it forever.**
- Your body sends clear signals regarding how it feels about your window; it's up to you to decode them.
- Your window impacts your hormones, including your Switch. Make sure to check in on them.

- A window of ten to twelve hours is better for most people most of the time, for two reasons. It gives the body sufficient time off. Taking in calories over ten to twelve hours gives the body twelve to fourteen hours to do its recovery work. Longer time-outs, such as sixteen to twenty hours of not eating, make it more challenging to get in all the nutrients your body needs in keeping better nutrition (balance, quantity, timing, quality). The window of ten to twelve hours also breaks up the day into pit stops at intervals of about three hours.

Caloric window + pit stops = better timing

Whether you wake up earlier than 6 a.m. or at varying times, use the caloric window calculator:

Waking time + up to two hours, ideally (not longer than three hours) = first pit stop

TEN-HOUR WINDOW	TWELVE-HOUR WINDOW
Pit stops: 8 a.m., 11:30 a.m., 3 p.m., 6 p.m.	Pit stops: 8 a.m., 11 a.m., 2 p.m., 5 p.m., 8 p.m.

Then about every three hours = a pit stop

Bedtime minus about three hours = last pit stop

If you answered yes to any of the eight statements in the timing part of the assessment, here's how you adjust the components to secure better timing most days. Make your first pit stop—better quantities of nutrients, in balance—within two hours of waking.

Example: Instead of black coffee with nothing to eat, have your coffee with Greek yogurt, berries, and hemp seeds or an open-faced egg and turkey bacon sandwich. If your coffee is a latte, then wait three hours for the egg and turkey bacon sandwich, and . . . make that sandwich your next pit stop. And yes, you can do coffee with a protein powder as a pre-workout option. Just remember the "Better Be Delicious to Me Right Now" tool. If this combo makes your coffee not delicious to you, there may be consequences with your post-workout choice.

If, like Sarah, you don't have "time for lunch" three hours later, have a nutrient-balanced pit stop—it can be a grab-n-go option such as a bar. Be sure to get in your next pit stop about three hours later. For that last pit stop, reverse engineer from your set bedtime. Make it everything you and your body need it to be to go the final three hours with no more than a noncaloric beverage such as water or herbal tea leading up to bed.

Practice better-timed choices to see what is doable and delivers better for your Switch and ecosystem. Evaluate the timing experiments with assessments of your daytime energy, sleep quality, cravings, body composition, and other data.

"But what if I have tried adjusting my timing and it didn't work?"

Do any of these sound familiar?

- I am not hungry every three hours.
- If I have breakfast, I am hungry all day.
- If I eat this often, I will eat too much and gain weight.
- I am really busy. My life doesn't allow for a meal every three hours.
- My ____ gets home late, and we like to eat together.

If the timing I'm suggesting hasn't historically worked for you, or feels like it won't, two things could be going on separately or simultaneously. Perfectionism may be at play. Remember, we're all about better, not perfect. Life doesn't happen in three-hour increments, so your pit stops may not always happen in this window. When they don't, how your body responds can give you insights about your body's needs and its responses to not getting resources in a consistent, timely manner. The idea is to experiment with eating every three-ish hours and see how your body responds. When a ten-to-twelve-hour nutrient window isn't doable, or when you're too hungry to wait three hours (or you feel too full to have anything), this gives us insights that other pillars or your digestion, for example, need attention.

The tenets of the timing pillar—trialing a ten-to-twelve-hour window and pit-stopping about every three hours—work because they are based on the human body's design. They don't work when the body's design is being challenged or is dysfunctional. That's why we worked on digestion and hydration first, and why timing is a pillar that works in cahoots with the others, not as a standalone.

Quality

A cast-iron pan is round, and so is a paper plate, but that doesn't mean they're equally useful for frying an egg. A car runs better on certain liquids—gasoline, ethanol—than it does on banana juice. You get an email in a script that does not even look like a language; just because this email reached your inbox doesn't mean you can read or answer it. When our bodies recognize the things they get, they can use those things—and use them more efficiently and effectively. What they don't recognize easily or fully they may be unable to use at all. And our bodies send us signals. Some signals register with us fast—give a kid candy and watch what happens. Other signals reach us on a delay—artificial sweeteners disturb our oral and gut microbiome, but we can be slow in becoming aware of the trouble this causes. Signals reacting to quality tend to be of the slow-impact kind. So where you are today with regard to quality is an important baseline. The signals you're getting right now are a cumulative story. Let's consider how the *quality* of your regular choices may be frustrating your body and giving it unproductive or wasteful work.

Consider the review you did to familiarize yourself with better and not-better quality ingredient sources. For the choices you consume most often and know the quality of: Were ingredients, preparation, or storage mostly on the "Not Better" side of the list?

If that's a yes, you have a great place to start on quality adjustments:

- Check out the quality of what you consume regularly.
- Upgrade where doable, as indicated.

Here's how we do that:

- Use both sides of the list. Look at the "better" column. How can these choices figure into your pit stops more often?
- Look at the not-better items for ones to substitute or reduce now or within thirty days.
- As you experiment with upgrading choices, which can feel cost-challenging or prohibitive, consider setting a weekly budget for "experiments." Use this budget to buy a new choice and see if it is delicious to you and works as a quality upgrade.

- Beyond the list, the online resources guide you to tools that help you identify and upgrade choices. These include third-party certifications, websites, and lists designed to navigate through the quality of choices. You'll get to know your own go-tos. Check them out via the QR code on page 327.

For most of us, the quality of a lot of what we consume may be unknown because someone else is preparing it:

- To get insights on the quality of ingredients, preparation, and storage of these regular choices, you can always go more directly to the source, reviewing restaurant, product, and ingredient websites or asking business owners and manufacturers directly.
- If for any reason you are not going to, consider reducing the frequency of having items for which you don't know the quality.

Before we move on, some words of caution and consideration. Plus, another story about fly-fishing.

Before my Alaskan fishing adventure, I got a primary lesson in quality from a guide in Montana. The guy had taken me out to teach me some basics of fly casting. What he knew about his customer: plant-based, gluten-free, and a dietitian. The morning went well. The fishing was catch and release, so I was curious about the menu for lunch. There was a picnicky-looking bag. Vegan in the boonies, I expected PB&J. I was wrong. Out of the bag came a ceramic bowl, a glass container with salad, and a cloth napkin. I watched wide-eyed like Jane and Michael Banks. What would my Mary Poppins produce next? He uncovered the ceramic bowl. Bison. He held up the bowl and a vegan energy bar. "What do you prefer? I know you don't eat meat, but I wanted to offer it since I hunted and cooked it myself. Or the bar. It's made with soy protein isolate," he deadpanned. "I don't eat that, so it's all yours if you want it." The bison was local, wild, almost for sure grass-fed and free-range. Never saw a feedlot or GMO corn in its life. It had lived on what was better for its body. Whereas the energy bar . . . My guide and host mentioned that he couldn't explain to his kids what it was made of, and he'd gotten a good grade in chemistry. The bison wasn't just the more appealing choice. It was the better-quality choice. The guy who'd prepared it and offered it to me had just shown me what I really was where food was concerned: a qualitarian. Been one ever since. Maybe you're one and haven't made that discovery yet.

That bison was delicious.

The quality calculus is sometimes, say, two kinds of chicken: battery-fed and raised with growth hormones and antibiotics over here; organic, natural-feed, no meds over there. More often, everyday life presents us choices between foods that are unalike. Maybe not between bison and an energy bar but two protein options on a menu, two choices at a breakfast meeting, two packaged snacks on a plane. Qualitarian living means you get to say yes to one choice way more than no to all choices, and that's great when it comes to adventures, culinary and otherwise.

Better quality does not necessarily mean eating only organic fruits and vegetables. Resources like the Environmental Working Group's "Dirty Dozen" and "Clean Fifteen" rightly get a lot of attention. They tell us about the fruits and vegetables that come to us with the heaviest pesticide residues. It's quality intel for the fruits and veggies we hopefully consume often, although not nearly as often as things like coffee, tea, chocolate, flour-based products, protein bars, and the like. Additionally, if the choice is between eating conventional fruits and vegetables or none at all, I come down in favor of nonorganic almost every time. But when it comes to these others? Better to pass. Try to hold your coffee (and the water it's brewed, dripped, or expressed with), chocolate, and protein bar to a better quality standard or find an alternative. The Environmental Working Group has some great resources to help you navigate these choices, too.

If something you love is not so lovable by quality measures, and you're not into giving it up or substituting something that's better, look at loving that thing less often. Maybe it's a treat or even a rare treat. When it comes to things that aren't helpful to our weight-health ecosystems and our weight health, it is very often the fact that less equals more.

It isn't better quality just because they tell you so. We're seeing an ongoing decline in trust toward food sources. The decline is mostly led by marketing. But even when swaying your purchase behavior isn't the desired outcome, in spite of good intentions, what passes some "better-quality" test often doesn't stand up. For example, that extra-virgin olive oil may have been adulterated with oils on the not-better side, especially in restaurant and food service. For a retail purchase, find a third-party certification that verifies it as both genuine (made from olives) and pure (from olives only). In a restaurant, ask to see the container the olive oil came in; this will give you a degree of quality assurance. Unfortunately, there are times when even that effort doesn't pan out—like my client who observed a restaurant worker refilling olive oil from a container with a different label. If that seems too dicey, based on your health and goals, you'll likely want to pursue your quest for better quality differently. Maybe bring your own or eat at home.

There's reasonable concern as to whether better quality is affordable. Better quality can cost more. But it doesn't have to. Typically the more people involved in the foods we consume, the more we pay for them. And of course, the more high-quality ingredients in a food's makeup, the more it's likely to cost (100 percent pure vanilla extract costs more than other kinds). However, we can invest our efforts and choose differently to make better quality affordable more of the time. Making a mostly organic salad at home can be half the cost of a conventional salad at a take-out place, and a third the price of a comparable organic salad in a restaurant. More economical yet: frozen organic vegetables sautéed instead of the same or similar in a ready-to-eat organic (often misleadingly labeled "fresh") package. Using coupons or buying bulk can help extend resources, as can expanding your repertoire of pit stops made from reimagined leftovers.

Better quality is pointless if it's not doable, and doable when you want or need to do it—which is to say, now. Whether what holds us back is cost, access, or straight-out non-deliciousness, quality upgrades don't amount to better if we're not able to make them more of the time or most of the time. Still, lower quality, due to these impediments, sure doesn't add up to better either. The aim is not perfect, just better.

The degree and time frame in which you make quality improvements—and notice outcomes—is a lot like counting steps. Yes, there is a mass recommendation for 10,000 steps daily. If you are at 9,500 today, that's a great goal. But if you are at 1,000, then maybe your better next steps will be to get to 1,200 daily for a while, then 1,500.

To summarize, when it comes to quality:

Nobody knows quality like your weight-health ecosystem.

A primary quality control officer is your Switch.

Better quality helps your body work better more often.

The choices you consume regularly need to give your body what it recognizes and can use most efficiently and effectively. That doesn't necessarily mean organic. It does mean those choices are full of nutrients with minimal challengers for your body to manage.

Do I need to detox?

Yes, every day that ends in *Y*, your whole life. Your weight-health ecosystem comes with a built-in detoxification function. It requires optimally resourced amounts of better-quality choices. It runs better when it can manage its workload (i.e., the challengers). So your goal with quality is to make choices more often that minimize toxins and help deliver nutrients to support detoxification efforts.

So my body runs better when I invest in the choices that give it fuel it can use while limiting the challenges I throw at it, too. And my wallet runs better when I invest in the choices I can afford. So I run better with both?

Exactly. You got it.

Better quality matters. But "perfect" or "best" quality is a myth.

Balance, quantity, timing, quality: Mix them together and, baby, you've got a sauce.

It better be delicious, for your body and for you right now.

But we don't eat pizza sauce by itself.

We put the sauce on the crust, add the cheese and the toppings, and then we fire the whole beautiful thing.

The components all work together.

All this talking about food and beverages . . . we better get moving, right?

Sprinkle or cover it with cheese.

ADD THE CHEESE: DUMPING "DON'T STRESS," "BE MINDFUL," AND OTHER UNHELPFUL LIFESTYLE ADVICE

Heading the list of things that aren't helpful and frankly piss me off are recommendations to just do this one specific thing; yet, if we could already do that thing, we wouldn't be seeking professional help. Like a cardiologist who advises "reduce stress" to improve heart health or a dietitian who tells you to "eat more mindfully." We are not going to do that here.

Here you choose your pizza's cheese. Cheese that will be delicious, doable, and personalized. Doesn't help if a pizza has a gooey cheesy layer but it sends you to the bathroom or makes your skin erupt. Same if you choose the plant-based cheese but it's not made from better-quality ingredients or it doesn't melt.

For our pizza, the cheese is the lifestyle choices your body needs for optimal Switch function. They happen for all of us— dairy or plant-based—but how much, how often, what kind, and in what balance is where we differ. Around my family table, there are some of us that add a dusting, more of us that enjoy a good amount of cheese, and a few who add so much they have no idea what is actually underneath. That's the same with lifestyle choices. Your cheese works with your sauce, and vice versa. Like a pizza, the whole is so much more than the sum of its parts.

THE FOUR PILLARS OF BETTER LIFESTYLE CHOICES

Turns out, to function optimally, the weight-health ecosystem also relies on lifestyle choices that align with the four pillars. It isn't just what you choose to do—or don't—but the balance, quantity, timing, and quality of those choices that result in your body getting what it needs when it needs it, more often. Now we will investigate the fundamental lifestyle choices—movement, sleep, stress, breathing, and joy—that impact Switch and ecosystem operations.

Let's kick off this section with the Big Three: movement, sleep, and stress. It's probably no surprise to hear that all three are hugely important to the functions of your ecosystem and your Switch. And as you probably can attest to firsthand, the fabric of modern-day life is built to screw with all three of these.

Diagnosis of and remedies for all three tend to add up to "This is a problem" with no solution offered. If I could address the problem, I wouldn't have the problem. So what is newsworthy here: approaches to optimize movement, sleep, and your stress response that are actually doable.

Movement

It's all calories in, calories out . . . that's why movement matters, right?

No, we unlearned that mantra.

Sitting is the new smoking . . . okay, we're getting closer.

But still no cigar.

Switch function impacts movement, *and* optimizing movement using the four pillars improves Switch and ecosystem function.

One form of movement we have already explored is digestive activity. Your digestive-tract muscles run throughout your body using movement to facilitate transport and elimination. Nurturing these muscles is a key part of Switch Optimization. The quality of your movement—where, when, how much, how intensely—will impact your digestion. The quality of your muscles—slow- and fast-twitch—will, too.

In order to move effectively and efficiently, your body also needs recovery. To not move or to move at a slower pace, in a less taxing manner, invites different ecosystem activity to occur—or not—with desirable outcomes. The quality, timing, and quantity of your recovery efforts impact your vagus nerve and thus the transport pace for your Switch hormones.

Then there is the movement that your Switch hormones directly impact. Organs—for example, the stomach—receive messaging for how they are to move—or not. Case in point, the prolonged intentional delay of gastric emptying accomplished by the Shot and more modestly with your Switch hormones.

In view of all this, we now turn our attention to decode the signals that your Switch hormones are revealing about your current movement choices.

Your movement assessment: Answer the following according

to last week—or if it was an abnormal week, use a prior week for evaluation. Answer yes or no and note the date with your answers.

Do you lift items or weights that feel heavy to you three or more days a week?

Do you walk or jog twenty minutes or more most days?

Do you complete more than thirty minutes of intense cardiovascular exercise two or more times a week?

Do you avoid being seated for longer than three hours (180 minutes) at a time during the day when not sleeping?

Do you spend more than twenty minutes outside in nature most days?

Do you stretch or do an activity that moves your abdomen more than once daily?

Are you able to do ten pushups? (On your knees counts.)

Do you go through your day with minimal—or earned—soreness from your activities the day before?

Do you fall asleep and stay asleep easily following your day's activity?

Do you spend more than fifteen minutes most days stretching or being stretched to get into your muscles and fascia for recovery? (Fascia? It's connective tissue that covers organs, muscles, bones—even nerves—and supports their movement within the ecosystem.)

Are your hunger and thirst after exercising satisfied with the choices you make in the hours following?

For any no answers, these represent areas of opportunity to experiment with improving the balance, quantity, timing, and quality of your movement.

While I am passionate about movement, exercise, form, and recovery, I am not a trained expert in any of them. What I have put together here is what I have learned through collaboration with experts to support my patients. This is how they've taught me to assess potential imbalances, and how to evaluate improvements or ongoing challenges. For building and refining plans, I suggest collaborating with an expert.

Balance

To move effectively, the muscles of your ecosystem work in opposition—push and pull, contract and relax. You have fast-twitch and slow-twitch muscles that require equal attention. More globally, the body needs movement to occur both intensely and moderately. And of course it needs you to both move and be still.

Nutrients play a key role in the push-pull, contract-relax efforts. We saw this in hydration with electrolyte minerals, where imbalances were as critical to identify as insufficiency of one or more nutrients. Activity choices should do the same.

Muscles should be used in both directions—contracting and relaxing, with pushing and pulling. Recall the fly-fishing guides hauling the boat anchors and all the gear as well as their new guest's kettlebell-containing duffel. What counts as your movement isn't just about what you do at a gym for minutes or an hour a few days a week or even daily. Muscle-working movements occur in cleaning and gardening, lifting children or groceries—even in pushing away from the table. For your Switch, you need to make sure this includes abdominal movement to support digestion along with strength-training efforts to build and retain the amount and integrity of muscle mass throughout your body.

Intensity of activity—in exercise, this is now commonly referred to as zone training; in old-school terms, "degree of difficulty"—should be balanced, too. Your personal training goals and lifestyle will impact this dramatically, so the term *balance* can apply quite differently to your life versus others' and throughout your own life phases or seasons. But balance is still key. What does this mean? The harder you train, the harder you need to recover. From professional athletes to entertainers to parents, caregivers, and professionals such as doctors, teachers, builders, and farmers, when I audit their "training" time, I can often find a severe imbalance with their recovery time and efforts. Often on their to-do lists is "I need to make time for the gym," when what their body is saying is "You need to make time to stretch and relax muscles, open up fascia, and recover."

We can use several markers beyond the movement assessment above to evaluate the balance of your movement. Continuous glucose monitoring tells me how your body is responding to training and where recovery is suboptimal. Your body's oxygen level, pulse, and respiratory rate give me insights; and if you have the ability to monitor heart rate variability, we get even more usable data. Scales and scans that measure body composition provide data, too. As does how you feel—physically overall, on waking, and throughout the day—in terms of energy.

Quantity
We've already broken up with the word *more*, so let's see what better quantity of movement locks like for you.

- Activity is like macronutrients in that we need an optimal amount of them all and, as we just covered, in balance. Think about movement types in this analogy: your cardiovascular exercise = movement carbs; strength training = movement protein; stretching, fascia, and recovery work = movement fats. Using the movement assessment, you can see where you may benefit from increasing the quantity of a certain type of movement or recovery. The specific amount will be dictated by what is doable in your schedule and how your body responds.

- Just like with nutrition, if there is a reason you need to limit or not do one form of movement, you will want to use the others and some supplemental sources to make up for what your body isn't getting. What does that look like? It is hard to move a lot when you are on a plane, but you can move your abdomen and other parts of your body in your seat, and you may be able to get up and walk around. Before you board or between flights you can walk around the airport. You may also benefit from wearing compression socks or using other tools to help your body when movement is impaired.

- Overenthusiasm or an active memory of glory days often leads to overexertion in the name of meeting a prescribed movement goal. If climbing the stadium stairs today means you are unable to move for three days afterward, your in-the-moment movement choice wasn't better. Temper enthusiasm with a small dose of reality by asking yourself—as we do in the assessment—how a movement undertaken today is likely to feel tomorrow. Doability like this is compatible with setting big movement goals. Do set them, train for them, learn from how your body responds, optimize recovery, and smash your big goals!

Timing

Just as with our consumption of nutrients and calories, the amount of movement we do can leave our ecosystem under-resourced or overwhelmed. Doing a lot of activity at just one point in the day and spending the rest sedentary isn't better. We need to break up movement into segments. That's right, we are about to talk about pit stops . . . for activity. The concept of exercise snacking is having its day, and I am here for it. But beyond "exercise," our body will respond even more favorably to movement snacks—pit stops at least every three hours where we move, and move differently in type or intensity, to stimulate and challenge the body to shift directions.

Using activity as your go-to stimulant over another cup of coffee or tea is a Switch ecosystem win. Too frequently we rely on caffeine or other stimulants in place of our own energy-producing efforts: movement. If you can't sit at your desk and do your work for more than three hours without another cup of joe, there is a valid reason. Your body doesn't want you to; it's screaming for recess!

Another timing factor is our "active window," similar to our caloric window. We want to be active when we are awake, at least every three hours, but we also want to make sure to give our body optimal, not just sufficient, rest. Think of it as a balance of time spent horizontal with time vertical. More on this in the sleep section where we focus on setting and sticking to a consistent bedtime.

Quality

Quality of movement often focuses on how we move to avoid injury and to support optimal outcomes. Here we expand the concept to also focus on *where* we move—the parts of the body and the place (like the gym versus nature).

- Remember how I mentioned working with an expert? One reason for doing this—or using a mirror or image tracker—is that you get feedback on your form. You want to move well, safely, and intensely enough to activate muscle synthesis and avoid injury.
- Abdominal movement is a must for building and maintaining strength but also to encourage your digestive muscles to work efficiently and effectively. Later on, we will look at how you can use abdominal movement to address motility challenges.
- Working your fascia, like working your muscles, maintains the tissue's integrity. Think of it like the lining of your digestive tract—it needs movement to keep it functioning optimally. You can work your fascia in both controlled movement, such as yoga or tai chi, and explosive movement, such as hopping, jumping, and skipping. Foam rolling, a type of stretching using anything from PVC pipe to foam cylinders, is a great tool for working out your fascia.
- Movement in nature provides support for your ecosystem, including stimulating your vagus nerve. Switch Optimization in effect! Want to really optimize? Go forest bathing. This isn't about hiking intensely in the woods but rather slowing down to allow your senses to experience the forest. Forest bathing has been shown to increase heart rate variability, a key performance indicator of how your body is resting.

Two Tales of Not Moving Better . . . Until They Did

My patient Fred was trying to optimize his weight composition and reported that he ran five miles on a very hilly path most days. He lifted weights twice a week. Yet his lean body and fat mass weren't changing. He asked what else he should get rid of (food and beverages). I noted that his continuous glucose monitor sensor showed minimal elevations in blood sugar despite an intense daily run. He was surprised when I told him that the blood sugar data validated a concern I had—that his body had acclimatized to his runs, essentially not finding them stressful (which I knew because his blood sugar was not going out of range), so the answer was actually to adjust his fitness, not his food. He worked with his trainer to make fitness adjustments, and it paid off. Later he sent me a note: "I had no idea my body was bored with my routine. Thank you."

A busy CEO and dad, Edward admitted that his time to exercise had really shrunk in the last few years. Whenever he could, he went to the gym and got in cardio on the treadmill. He ran fast for thirty minutes, and if there was time, he added some weights. He was forty years old, and his weight composition and health goals shined a light on a need to build and protect his lean body mass. He had a thin, athletic build, and his genetics revealed that his body should respond well to more intense training. I suggested he pivot his priorities to thirty minutes of strength training and then add in activity over the day such as stairs, squats in place, or walking meetings instead of seated ones. In three months, he saw a gain in lean body mass along with a reduction in his cholesterol. Switching the balance of his movement, like adjusting macronutrients, made the difference.

Sleep

For many of us, at various points in our lives, two of the biggest issues for our Switch and ecosystem—caloric window and sleep window—wind up in a codependent relationship. When we finish our calories earlier and go to bed at a better time, we experience weight-health harmony. This section is like couples therapy to identify and resolve the problem. A *consistent* bedtime is hugely helpful. Two sleep experts I especially admire—Todd Anderson, a sleep performance expert; and Dr. Michael Breus, "the Sleep Doctor"—share compelling evidence that we need to bring as much vigor and discipline to sleep training as to any other training plan. Topping their lists: adopting set bed- and waking times.

Todd grabbed my attention presenting his concept of "social jet lag" to a room full of jet-lagged conference participants. Most of us undergo this kind of lag often, without crossing any time zones. We accumulate sleep debt during the week, then try to pay it down by getting more sleep over the weekend, and/or the weekend comes and we stay up late. Social jet lag exerts drag on a quarter to a third of our days and nights, a hundred to 150 days a year. "Your sleep window needs to be the same, seven days a week," Todd said.

Variations of even thirty minutes negatively impact your eco-system. That gives you a target when you're figuring out what's realistically doable. Your body needs sleep, the yin to movement's yang, to accomplish several key goals. For your body, the activity of being inactive is hugely important. We don't tend to treat it that way. For most, seven to eight hours each night is the quantity that will support enough recovery and the body's housekeeping duties. When we get too little sleep, we don't experience the different phases of sleep. Sleep timing is a lot like nutrient timing. Beware backloading or waiting too long to get it all in. An early bedtime is better, if doable, even if that means you have to get up to do work or address other items on your to-do list in the morning.

Social jet lag and sleep disturbances have Switch and ecosystem impacts. When fatigued, the body will call out for energy more frequently and intensely than when well rested. It accepts stand-ins, and they happen to exist everywhere in the form of refined carbohydrates and stimulants. The body keeps the score, as brilliantly noted by the psychiatrist and trauma researcher Bessel van der Kolk, so we need to proactively train for a better sleep window, a consistent one, for Switch and ecosystem optimization.

We practice sleep training in the first years of our lives and then spend the next decades breaking up with that training. As adults, we try to regain "better sleep," but until more recently that hasn't been with revisiting sleep training. Typically I hear from patients that they tried to address their sleep first with a supplement such as magnesium, then melatonin, then added medications. Or for some it was so bad right off the bat they started with medication, only to struggle with needing more or not being able to come off it to improve sleep. To add insult to injury, the medications often left them feeling less than well rested on waking.

Finally, sleep quality matters. Waking up to pee is not just annoying but also an interruption that may impede the body from getting through its recovery workload at night. The body wants to go through different phases of sleep where it sees to different aspects of its recovery work—think of it like a quick cleanup, a more thorough one, and a deep clean—to get to the recesses of what it has accumulated or is left to work on later. The recommended seven to eight hours is even better when it starts with the body getting fewer things to do earlier (stopping calories a few hours before bedtime, closing down screen time an hour or so before, and even changing light and temperature exposure).

Here are some keys to sleep training. Think of them as experiments and track how you feel after implementing any of them for a few nights.

- Set and keep a consistent bedtime, even on weekends. This may mean that you don't go to bed earlier if it would not be sustainable for the whole week. That said, this isn't permission to just set a later bedtime. Evaluate where you can adjust schedules—dinnertime, what you do after dinner, travel schedules, and social plans. Remember: better, not perfect; but also keep in mind accountability and doability.
- Set and keep a better caloric window. Concluding your calorie intake early in the evening has multiple benefits, including better sleep.
- Establish a sleeping room—for sleep and sex, and that's it. Where possible, remove chargers and entertainment options (including books and other reading matter), and minimize light and sound challenges. If you have minimal control over the environment, consider a sleep mask and ear plugs.
- Take breaks—naps, breathing, and meditation moments. Pauses help the body experience recovery when life happens. Naps do not "count" toward your sleep hours, even if your device adds them up, but they do count toward the body shifting into rest-and-digest mode to give it an opportunity to do these.

Sleep training is worth your investment, just like fitness training. If you need some extra support, seek out an expert who will get to know you, your body, and your goals to personalize your sleep plan.

Stress

Far from a design flaw, our stress response alerts us and helps us focus on perceived danger and performance opportunities. Like your Switch-hormone activity and blood sugar levels, stress elevation is meant to occur in response to stimuli and then get the signal to turn off and return to baseline. So, we've actually got a pretty smart response system—when the building's on fire, we don't need to worry about signaling time for lunch. The stress response both impacts the Switch and is impacted by it. For Switch messages to travel effectively via the vagus nerve, the stress response needs to be dialed down. When activated, stress hampers your digestive efforts to keep things moving, thus challenging Switch function.

Nutrients play one role in the regulation of the stress response, with magnesium and calcium as its primary gatekeepers inside and outside the cell. Non-nutrient stress efforts are the yin to their yang, so let's go there now with a three-pronged approach to optimization that works to ensure stress isn't negatively impacting your vagus nerve and motility, and as a result, your Switch function.

Identify **elevated stress** and **reduce** it.

Breathe better more often.

Make choices that generate **joy**.

You don't have to nor should you attempt to remove all stress from your life. It turns out that's impossible. A good balance of stress helps us meet our goals and even function *better*.

Elevated stress is the situation we need to identify. It is during elevated stress that the body deprioritizes motility in favor of a hyperfocus on efforts to address the stressor. Today, we have numerous personal diagnostic tools to evaluate elevated stress (labs, apps, watches and other devices to self-check blood pressure, wearables such as a CGM or ring, etc.). The most common result: too high. The most common prescription: reduce. Neither of those is particularly actionable, especially in a high-stress moment. Always on the actionability trail, I assess chronic elevated stress in investigating ecosystem disturbances. It turns out that reducing chronic elevated stress is a specific, measurable, and achievable goal. The tool I built, the "stress check-in," includes your assessment, experimentation, and optimization. The application starts in observation mode, since initially, due to stress, we lack the composure to de-stress enough to focus.

ASSESSMENT: HOW THE BODY TRIES TO TELL YOU STRESS IS ELEVATED

- Anxiety
- Elevated heart rate
- Cold, clammy skin
- Bloating, constipation
- Loose stools, diarrhea
- Racing thoughts
- Inability to focus
- Irritated eyes
- Difficulty breathing
- Breathing from your upper chest/shoulders versus belly/ full body
- Poor sleep
- Indigestion/heartburn

- Sugar/sweet cravings
- Exhaustion (despite having slept)
- Forgetfulness, brain fog

Experimentation: When one or more of these occur, you can do a stress check-in as outlined below. However, if you are new to doing stress check-ins or if you feel like any or several of the above occur often, it can feel stressful to figure out when to do the check-ins. It may be easier to choose three times in your day to check in on your stress. I suggest the following for ease in doing the check-in consistently, so you can gauge improvements; feel free to adjust your check-in schedule based on your body's signals and when you feel comfortable doing the check-in.

- On waking, when you are brushing your teeth
- A midafternoon point either before lunch or when you may experience energy shifts, such as 2–3 p.m.
- At bedtime, when you are brushing your teeth

Do three stress check-ins daily feel *stressful*?

A starter stress assessment includes one check-in at any point in your day. Ideally you choose a time when you are suspecting elevated stress or your body is sending you a stressed-out signal.

How to Do a Stress Check-In

1. Give yourself five minutes. Set aside all the things you're doing. If you can't take five, three minutes will do.
2. Rate and evaluate your stress.
   a. On a scale of 1–10, where 1 is the least amount of stress and 10 is the highest, where would you rate your current stress?

b. Results

    i. If your stress check-in is greater than a 6, it's essential that you *stop* and address your stress using the optimization methods that follow, until it is a 6 or less.

    ii. If your stress check-in revealed that you are a 4, 5, or 6, it's a good idea that you take action as provided in the optimizations to address your stress until it is less than a 4.

    iii. If your body is still sending you any signals (from the list at the top of assessment) that it is in an elevated stress state, it's a good idea to proceed with addressing your stress until the signal/symptom lessens.

Optimization: Reducing Elevated Stress

Choose one stress-addressing action based on what feels doable and likely to make you feel more relaxed in the moment. Don't stress over the choice. Figuring out which works better for certain scenarios is part of the elevated stress check-in process. There is no wrong choice, and there are likely choices that will work for you that are not listed below. Here are some examples:

1. Let it go away.

- Experiment with 3–5 minutes to clear your mind with meditation, where you sit comfortably, close your eyes, breathe better (see "Blow it away" below), and focus on your breathing instead of any thoughts.
- You may want to use guided meditation via an app or a recording to help you get started.

2. Laugh it away.

- Experiment with 60 seconds of sustained, authentic laughing.

Think about something that makes you laugh, grab an image, or listen to something that helps you laugh.

3. Blow it away.
- Start inhaling and exhaling, with a goal of each breath having a longer exhale than inhale. See if your exhale can become twice as long as your inhale. Do at least ten breath cycles.
- Close your mouth and inhale through your nose for a count of 4, hold for a count of 7, open your mouth and exhale emphatically for a count of 8. Repeat at least ten cycles. I first learned this 4–7–8 breathing technique from Dr. Andrew Weil and it's been a game changer for me and my patients.
- Breathe in through your nose into your belly until it is "full," then exhale saying "Ha." Repeat at least ten cycles.

4. HIIT (High-Intensity Interval Training) it away.
- Do a 5-minute series such as 20 seconds of jumping jacks or high-knee marching, 10 seconds of rest, 20 seconds of side-to-side jumping with feet together, 10 seconds of rest, 20 seconds of push-ups, 10 seconds of rest, 20 seconds of bear crawl, 10 seconds of rest.
- Do 30 seconds of jumping rope (with or without rope), 10 seconds of rest, 20 seconds of burpees, 10 seconds rest. Repeat 5 times.

5. Stretch/roll it away.
- Do a 5-minute foam roll of your whole body and use a hard ball (like lacrosse or baseball) on your feet.
- Do a 10-minute whole-body stretch.

6. Wash it away.

- Immerse in an Epsom salt or magnesium bath (not hot, but warm) for 10 minutes.
- Take a warm shower, finish with 30 seconds of very cold water.

7. Smell it away.

- Use essential oils or fresh lemon or herbs to deeply inhale through your nose (and allow for a longer exhale through your nose or mouth—see "Blow it away" above).

8. Send it away.

- Get out into nature—ideally touch nature (hug a tree, remove shoes and socks and walk on the earth) for 10 minutes.
- Spend time with an animal outside.
- Color, draw, paint, or take photos outside (okay to post on social media later).

Take a Moment to Reassess Your Stress

Did the stress-addressing activity you chose reduce your elevated stress to a better level (a 6 or less)? Over time, you may notice that certain activities address certain stressors better than others. Use this information to direct your future choices when you experience elevated stress.

Essentially we want to learn to decode the body's elevated stress signals and to experiment with stress-reduction techniques, noting that some will work better than others based on the type and timing of stress. For example, if you are on the toilet attempting a bowel movement and you experience stress because it is difficult, specific breathing exercises may be a good tool. During an incident where you are frustrated by someone's remarks or inactions, exiting the interaction may be a more effective stress reduction tool—and so can full belly laughter. This last example also reveals how many times the sources of our stress are ones over which we have little or no control. Reducing such sources of stress as much as possible is within your control.

Two ways to reduce elevated stress that deserve our attention are **breathing** and **joy**. Let's move on to them now.

Breathing

It is a requirement for your Switch and ecosystem (and, um, living) that you are almost always breathing. But how you breathe will impact their function considerably. As a society, for multiple reasons, we've shifted to being poor breathers. Talented experts—and assessment tools, both free DIY and wearables—have brought attention to the epidemic of poor breathing and its impact on our health span. We are relearning the value of nose breathing (thanks, James Nestor, for the powerful book *Breath*) and exploring tools such as mouth taping and self-assessment. As with movement, I am not the expert here, but I am a convert for myself and my patients. Optimizing your breathing is invaluable—and free. It may also take you into the realm of oral cavity assessment and even dental work to optimize your ability to breathe better. (*Note:* It did for me, and I had never been more surprised to learn—nor had I *ever* been told—that I have a very small mouth! Targeted adjustments improved my oral health, which improved my digestive health and sleep health.)

As with hydration, the intake of oxygen alone doesn't confer health benefits. Oxygen doesn't just need to get to our lungs but to several other places in our bodies. After taking it in, we need to move oxygen around, to get this vital element where it's needed, and to eliminate its cellular waste product, carbon dioxide. When we breathe, the diaphragm contracts and massages what is underneath—the abdominal organs—which promotes motility. Better breathing also stimulates the vagus nerve to shift from fight-or-flight mode into rest-and-digest mode, where it runs communications optimally. Notably, then, breathing is a tool to optimize your Switch so it receives and delivers messages most effectively and efficiently.

Experiment with better breathing. Here are more details on two exercises to start with—even right now. It will give you a head start on what's to come next in the assessment.

- Place one hand on your belly and one hand on your chest. As you inhale, over about four seconds, work your breathing so you feel the movement in your belly more than in your chest. Hold your breath for two or so seconds, then exhale, similarly working to feel your exhale occur in your stomach area.

  A trick to make this easier and more effective: Wrap a tape measure or a string with a marked spot that you put over your belly button around your midsection. Do the breathing exercise above. See whether your midsection expanded about an inch when you inhaled.
- 4–7–8 breathing. As introduced above, this breathing technique delivers big wins. I was blown away that there is a type of breathing that could compete favorably with blood pressure medications. Guidance on how to do this and other better-breathing resources are included in the online materials found using the QR code in Part VI.

Joy

Joy is, among other things, a tool that helps us build better health span. Not just the outcome of our efforts at being healthier but also a means to getting there. And it's the strongest antidote to INFObesity. When we're enjoying our lives, we're less susceptible to the endless loads of content barraging us from so many directions. Joy, as I experience it and hear others' relate their experiences, is different from happiness mainly in that it's something we build within ourselves and can access throughout our years. Joy is not resource dependent. It doesn't have to be rationed. We each control our own supply. And it's free.

Why do some people merely survive, while others go on to thrive? Although research struggles to pinpoint the reasons, one point that gets highlighted often is the resilience that a person has or develops in the face of challenges. Joy is a valuable asset for cultivating and bolstering resilience.

Research has shown the link between joy and various health benefits:

- **Vagus nerve function:** Joy happens on the non-stressed side of things, away from the flight or fight.
- **Heart health:** Joy lowers stress, lowers blood pressure, and improves overall heart function. Laughing plays a major role.
- **Immune system:** Joy and positive emotions correlate with higher levels of antibodies and better immune functions.
- **Stress:** Production of stress hormones such as cortisol decreases in a joyous, positive mood, rebalancing one's mental state.
- **Pain management:** Laughter and joy trigger the release of endorphins, which are natural pain relievers.

Whether prompted or on their own, my patients typically tie "wanting to experience more joy" in their lives as a driver or the driver for their immediate health goals. Many talk about family and social connections, and joy often comes from social interactions and building relationships. These connections are crucial for emotional support and can lead to a healthier, happier life and bring more joy. It's a positive feedback loop—chicken versus egg. Joy leads to more joy. But how do we get some going in the first place?

Start with the experiment below to help you find your current joy status and to understand how your current choices impact your joy.

The Better Joy Experiment

Joy operates as an input and as an outcome, very much per our four pillars:

Balance, Quantity, Timing, Quality

Let's take them out of order.

Do we need to be joyful all the time? No and yes. Doing or being anything 100 percent of the time is probably a terrible idea, not to mention impossible. We can skip the stress of struggling to achieve the unachievable. The result: anti-joy. Our lives have plenty of that already.

And yet . . . two things:

1. Even if you're not experiencing in-the-moment joy, if joy is a steady part of your life, it's there in the background while you grocery shop, wait for a car repair to be done, miss a flight,

hear about the whopping vet bill. It's that music that only you get to hear.

2. At many points in your life there are joy gaps. You can look at it two ways: (1) how much joy you experience versus how much joy would support your optimal health, and (2) how much joy you're getting from an hour or a day or an experience versus how much joy might be there for you.

Think about those gaps. Experiment to see if there are opportunities to increase the amount you already experience or to create joy at different times and how that makes your body and being feel. *More* than what? How much is more? If you don't know, your first step is to quantify your current joy, noting when and for how long you currently experience joy. That's the general, across-time part of the experiment.

Now, over the next few days:

- **Identify joy cues:** Take a moment to define how you will know you are experiencing joy. We call these "joy cues," and you will use these to mark the start and end of your joy. (Examples: when I laugh, when I feel fully present, when I start with a positive thought or attitude versus a negative or judgmental one . . .)
- **Track it:** Over these few days, note when you find yourself experiencing joy—the starting and ending points. (Example: 6 p.m. to 8 p.m.; that two-hour dinner with friends.)
- **Observe joy triggers:** What sparked your moments of joy? (Examples: being in the company of friends, yoga, taking a walk with your dog, meditation, laughter, waking up.)
- **Note joy interruptions:** Identify what suppressed, delayed, or

outright destroyed your joy. (Examples: a friend commenting on your weight, seeing an email notification, seeing something irritating or upsetting on TikTok.)

Evaluate your results:

- **Balance I:** Is the amount of time you are experiencing joy greater than the amount of time you are not? If no, this would be a better goal to work toward for most days.
- **Balance II:** Are there times of the day or days in the week that are joyless or when joy is less frequent? If yes, explore what could help you invite joy into these moments some of the time.
- **Quality I:** Look at what ignites or excites or stimulates your joy. Are these outward sources where your joy comes from or are they catalysts to joy that's inside you? Maybe that joy is tap-into-able with outer catalysts, or even without them?
- **Quality II:** Look at what triggers tend to move you from joyful to fearful or otherwise disrupt your joy. For each of these, think of ways that you could hold on to your joy for longer, reducing or removing the impact and frequency of these triggers.

What Is a Better Quantity of Joy?

As far as I know, no one has invented a joy meter with a set of metrics. And I'm good with that. Another gadget we can do without. We all are pretty skilled at keeping track of our joy to the extent that serves us. We know when we have more and when we have less. We can track our joy directionally without measuring it in units. We're also good at feeling different kinds of joy—different types and qualities.

Recognizing different kinds of joy and then recognizing the kinds that speak to you is a fulcrum for upping your joy quantities.

- **Use inspiration, on repeat:** My friend's dad loved telling me about his morning routine. On waking, he went downstairs and watched the last five minutes of the movie *Rudy*, which brought him pure joy. I witnessed it more than once—it was awesome! The "what" here is personal, but the focus on repeat for inspiration is the key. The "it" could be a letter from a friend, a poem, a quote you get as a tattoo or a screensaver.
- **Spend time with animals:** Dog owners report higher levels of joy. Dogs and other animals provide companionship, unconditional love, and a sense of purpose, all of which can contribute to experiencing joy. Plus, the simple act of petting or playing with your animals can release feel-good hormones such as oxytocin and serotonin. One of my patients tells me about a photo of her dog that she uses as a calm point in tense work moments and online meetings.
- **Get fresh air:** Getting out into nature can increase the production of endorphins, neurotransmitters, and a protein called brain-derived neurotrophic factor (BDNF) that helps brain growth, repair, and development. Taking a few moments to appreciate the beauty of nature can bring better health to you mentally and physically.
- **Get a visual cue from art or a cute picture** when you are using your digital devices to invite joy. A client used AI to create an image that is her computer wallpaper and invites joy for her.
- **Set boundaries and say no** (joyfully). Opting out of activities

or making choices to prioritize yourself can increase joy. Conversely, choosing to spend time connecting with someone or a set of people with whom you experience joy, by saying no to other commitments, can also be a way to improve your quantity of joy. Not compromising your needs is a strong strategic move to increase the amount of joy in your life. *Note:* This requires a step beyond saying no; you need to mean it. Saying no with conviction—truly choosing to miss out—is a way to experience joy: From FOMO, the fear of missing out, to JOMO, the joy of missing out, we like to say. If you say no and don't mean it—for example, you choose to stay home but then watch the "activity" via social media or texting with someone there, and question your choice—that doesn't invite more joy.

MELTING THE CHEESE

These lifestyle components—the cheese—as variable as they are, have in common being either Switch-engaging or Switch-enhancing. Many are both. This last section may feel like an overwhelming fondue of musts, shoulds, and not-doables. If so, stop. We can't have our cheese give us elevated stress. Recall that cheese can be a dusting, an accessory, a lot or a little, or a snowfall. But what it should always be is delicious for you. Use the table below to get a look at the overall recommendations, and the upcoming assessment will help you determine with which and how much you might experiment to add your cheese.

YOUR LIFESTYLE RECOMMENDATIONS (DAILY)	
QUANTITY	Activity *and* recovery daily guidelines: Focused exercise: >30 minutes most days; alternate types (see balance) and duration (see timing) Stretching: >15 minutes most days; alternate types (see balance) Movement: every 180 minutes for at least 5 minutes Sleep: >7 hours Playtime: >1 hour Screen time: <6 hours* Unlimited (but at least 30 minutes) laughter, time in nature, pauses, pets/touching, hugs, joy, and alone time *All screens combined.
BALANCE	Aim for diversity in the sources for each of the above (strength vs. cardio, higher vs. lower intensity, upper vs. lower vs. whole-body movements, types of sleep [REM, deep, etc.], recovery work, types of stretching [work out your fascia, etc.]). Choose exercise that challenges your body—it should feel uncomfortable or "hard"—at least twice weekly. Hard and uncomfortable does not mean painful. Get insights from someone who evaluates your body, especially if you have a current or past history of injury. Add different types of breathing, especially to remedy elevated stress. Balance time with others and time alone.

YOUR LIFESTYLE RECOMMENDATIONS (DAILY)	
TIMING	Plan pit stops for movement and mindset shifts about every 3 hours. Establish and try to maintain a consistent sleep window. Be in bed, attempting to sleep by 10 p.m.* Sleep and wake at the same time, weekday or weekend. Limit all screen time to begin at least an hour after waking and finish an hour before bed. Experiment with adding joy where it is not present *Or choose 1 hour earlier than you currently are getting into bed to fall asleep.
QUALITY	Stop caloric intake (that includes alcohol) 3 hours (at least) before bed. Create a cold, dark room for sleep. Experiment with breathing through your nose and with longer exhales than inhales. Attempt to reduce activities and thoughts that elevate stress (especially about things over which you have no or minimal control). Move your body in different ways. Avoid movements that create pain and repetition of the exact same movement types. Choose playtime activities that generate joy, happiness, soul-singing.

PART IV

HOW'S YOUR SWITCH FUNCTION?

But, Ashley, what's better for me? Recall how we started the pizza-building process with the description and delivery of a tool—the "Better Be Delicious to Me Right Now" test. It's time to apply that test to the pizza you've been building, because to be better for you, it has to be delicious to you right now—a metaphor here for matching what your body needs right now.

The flavors, textures, and aromas of a pizza that hits the spot are the result of finding the right ingredients, assembling them in the right order, firing at the right temperature, and serving the pizza to a body that wants *exactly* that version of the recipe. So far, we have the ingredients and the order, but for toppings and baking temperature, we need some critical information on how your Switch is functioning and what that means for your ecosystem. Let's (finally!) move on to the Switch-Function Assessment to map your Switch-function status, and then use that data to identify your better choices for optimization.

The Switch-Function Assessment happens in two phases:

Phase 1: Is yours optimal or suboptimal?

Phase 2: If suboptimal, is it suppressed, delayed, or dysfunctional?

To optimize your Switch function, you need to find out where your Switch stands right now.

That's what this—the first-ever Switch-Function Assessment—enables you to do. This assessment uses a combination of tools:

clinically validated, quantitative ones; and subjective, experiential ones. The assessment, as a resource that hasn't existed before, is evolving. My patients have contributed to it significantly, and I've now formulated it so that you can do it on your own.

Beyond establishing whether your Switch function is suboptimal, this assessment has four main objectives:

1. Establish benchmarks as starting points to enable you to see more clearly what happens when you implement new choices. You're looking for *directional improvement*, meaning improvement relative to your starting point. Such starting points have a way of motivating us. When it feels like we're not making progress, or at least not noticeable progress, a look back to where we've come from usually tells us that we are.

2. Combine a set of widely different parameters to inform your picture of where your Switch is, optimal or not; and if suboptimal, where it is on the continuum right now.

3. "Gaming" the mission to optimize and restore delivers a strong productive benefit. It's often helpful to get competitive—with ourselves.

4. Last but far from least: As you probably don't need to be told, the interactions that drive our weight-health ecosystem are complex. So engage the assessment not just for a thumbs-up or thumbs-down on the question of Switch-function status. The assessment is a mini-mission. It's the reconnaissance in which you do a lot of initial fact-finding and intelligence-gathering to get more specific and granular in personalizing *your* mission to optimize and restore weight-health hormone and ecosystem function.

We're going to do some GPS-type work and locate your Switch initially on this continuum:

SUBOPTIMAL　→　OPTIMAL

If your Switch function is not full-on optimal, we'll go to the next level.

To what degree and in what way is your Switch suboptimal? We get more specific on the suboptimal end of that spectrum:

DYSFUNCTIONAL　→　DELAYED　→　SUPPRESSED

Mission success is moving your Switch from where it is *toward* optimal function. That can occur in leaps from dysfunctional to

---

GPS ALERT

YES, THE ASSESSMENT IS NOT NEUTRAL

The Switch-Function Assessment is not geared toward confirming that Switch function is optimal.

It exists to help people zero in on their degree of suboptimal Switch function and make it their mission to address.

I can count on a hand and a half the number of patients who've come to me with optimal Switches. After all, the data show that 93 percent of Americans are metabolically unhealthy.

The vast majority of our Switches—nearly all of them—are struggling to keep up. And everybody's Switch at some point will experience challenges. Modern living ensures that.

optimal. More often it moves along the continuum as your choices initiate Switch-function improvements.

The Switch-Function Assessment rolls in these stages: (1) **gather and collect,** (2) **the Switch-Function Quiz,** and (3) **process your results.**

CHAPTER 9

THE SWITCH-FUNCTION ASSESSMENT

AN OVERVIEW

The Switch-Function Assessment is not a one and done. You are doing it first to find out where you are. Then you will reassess after about thirty days and continue to optimize. You will keep reassessing because Switch function for optimal weight health is a dynamic, lifelong process. So, you need a place to easily access your current and past data for comparison. To complete the assessment, choose whatever feels easiest and most reliable for you, personally. You have two options. You can complete the assessment online or you can use a notebook or a digital note-taking tool to complete this assessment. You will receive an email with your answers and details about how to process your results. To access the assessment online, follow the QR code that follows.

Stage 1: Gather and Collect

- Gather: This pulls in data you've likely already got.
- Collect: This pulls in data that you can easily obtain on your own or maybe with a bit of help. As you read through, you will see that all of this can be done on your own, but some of it may feel easier with someone to help you.

TIME COMMITMENT: It will depend on what data you have and how easily accessible your data is, but in general it takes about thirty minutes to an hour on one day, plus a few minutes daily over the course of a week. It's useful to first read through to get a timetable together. If you decide to enlist a person to help, get on their schedule.

Stage 2: Take the Switch-Function Quiz

With your gathered and collected data in hand, take the quiz.

TIME COMMITMENT: One block of about five minutes the day of your choosing, as long as you have all of your data in hand.

Stage 3: Process Your Results

Follow the instructions to interpret your quiz answers and learn your Switch function's current status: optimal or suboptimal; if suboptimal, then dysfunctional, delayed, or suppressed.

TIME COMMITMENT: Thirty minutes.
Let's get doing . . .

SWITCH-FUNCTION ASSESSMENT, STAGE 1: GATHER AND COLLECT

Gather Data

Earlier—for the crust—you did the first part of your digestive assessment and the hydration experiment. Gather that data now, if it is not already where you are collecting your Switch Optimization data.

Your next step is to gather additional data—current labs—as instructed below. These labs are important to the assessment and everything going forward in your personalization. I suggest focusing on them first, so you have a plan to have them in hand when you get to the quiz.

Instructions for lab data:

- Recent: You want close-to-current lab data, from within the past six months. If you've had lab work done in the past six months, have a look at the results. Some or all of these tests are likely there.
- For any you do not have currently, you may be able to get all or some (especially the core labs) done via your practitioner and insurance; you may also choose to invest in getting some or all of them on your own; the online resources will guide you to options.
  - ⟡ Core labs: hsCRP, hbA1C (also known as A1C), low-density lipoproteins (LDL), triglycerides (TGs), vitamin D 25 OH
  - ⟡ Additional labs: fasting insulin (not the same as fasting glucose), uric acid, glycated serum protein, apolipoprotein B (ApoB), aspartate aminotransferase (AST), alanine aminotransferase (ALT)

✧ Optional: a wearable continuous glucose monitor (CGM) sensor to evaluate blood sugar activity in response to your choices. If you use a CGM, gather the most recent data (this past week or month, for example) for time in range (for persons who are not diabetic, use 70–120 as your range; for diabetics, use 70–150 as your range), average glucose, and number of spikes and dips. Note this data for use in the upcoming quiz.

That's the lab data.

Collect Data

Body Composition Data Collection

Like your digestion and hydration, these data are key performance indicators of weight-health-ecosystem function. You've unlearned valuing the total number on the scale. What we are now collecting—your fat mass (the types and locations), skeletal muscle mass, and bone mass—helps you take your best shot and hit the bull's-eye as you personalize your optimization.

There are two ways to collect this data—by machine or by hand. The machine will give you location and type specific data in pounds (or kilos) and percentages. (Those done by hand, waist circumference—in the old days, "waistline"—or waist-to-hip ratio, will not. The latter give you only an approximation of your current body composition. However, that data is sufficient for our assessment purposes.)

• Do these ideally on waking, after a bowel movement and/or urinating but before eating or doing any exercise.
• Choose a day after a night that was as close to normal for you

as possible. (For those menstruating, better to check on days 5–15 of your cycle.)

By machine: Collect the data from the last ninety days via a DEXA scan, a bioelectrical impedance analysis (BIA) measuring scale, or from an app that does a visual body scan. While you may want to track ongoing progress in pounds or kilos (of body fat and skeletal muscle mass change), for the quiz, collect body fat percentage and skeletal muscle mass or lean (fat-free) body mass depending on what is provided. Also collect visceral fat mass, if available. There is no "best" scale for this despite the marketing claims. Ideally the one you use now will be accessible for future assessment.

By hand: If you don't have access to a scale or scan that provides a breakdown of body composition (fat mass by type, skeletal muscle mass, bone mass), you can work with two very good measures: waist circumference and waist-to-hip ratio.

- For this next set, measure in centimeters or inches; just use the same units for both.
- When measuring, you may want a partner, or it may be helpful to have a set mark on your body (one you make or that exists and doesn't move) to know where you measure each time.

A. Waist circumference (WC)

- Take this measurement three times and then go with the average.

- The steps:
  - ✧ Stand up. On each side of your torso, locate the tops of your hip bones and the bottom of your rib cage.
  - ✧ Lift up your shirt or top, or take it off, so you can measure directly on your body. Your clothes will add artificially to the measurement.
  - ✧ Halfway between your rib cage and hip bones, the narrowest part of your midsection, run a tape measure. (If you don't have a tape measure, a string works fine; you just need a yardstick or ruler to measure the string.)
  - ✧ The tape measure should cross your belly button.
  - ✧ Let out a breath. A normal exhale.
  - ✧ Pull the tape measure around your waist so it's snug but not tight. You don't want it to dig into your midriff.
  - ✧ Double-check that the measuring tape or string is parallel to the floor or the ground.
  - ✧ Note the number where the tape measure reconnects with itself. (If you're doing the string method, do the same— just mark the string and then measure it on a yardstick or ruler.) No need for fractions. To the even inch or centimeter is fine.
  - ✧ Repeat two more times. Take the average of the three measurements. (Add the three and divide by three.)

Record that number, the date, and the time; label it "WC" or whatever tags this measurement clearly.

B. **Waist-to-hip ratio (WHR)**

- Note your waist circumference average number (above).

- Now measure your hips. In contrast to your waist, which is meant to be the smallest area around your body, the hip area is considered the widest—specifically the widest part of your buttocks.
- Draw the tape measure or string around the widest part of your hips. For most of us, that's also where our butt sticks out the most.
- Take the measurement three times and compute the average. Use this number.
- To get your waist-to-hip ratio, divide your **waist circumference** (the number you got in A) by your **hip measurement** (the number you just calculated).

Record this number, today's date, and the time; label it "WHR."

Now it's time to exhale.

Breathing and Heart Rate Data Collection

All of these tests require a timer. Your timer can be digital, an old-fashioned watch that ticks, or a stopwatch.

The body oxygen level test (BOLT) is a measurement of the time it takes for your body to react to having the air supply shut off. (It's not a test of how long you can hold your breath.) It is a test that will provide insights about a key aspect of your Switch function. Do the BOLT now; don't rely on data from a previous test.

Step 1: Prepare

- My colleagues recommend taking the test first thing in the day. If it's not the first thing in the day now, go ahead anyway. You can do another test later.

- You will be timing the period between when you start to hold your breath and the time when you first feel the urge to breathe—the first stress in your body saying "Get air." These message-sensations may include the impulse to swallow; the constriction of your airways; and/or contractions of your breathing muscles in your abdomen, chest, or throat.

Step 2: The BOLT Test Prep

- Rest. To do an accurate BOLT, rest for 10 minutes before taking the test. While you do that, read through the following test instructions.
- After a 10-minute rest, do some belly breathing. Recall the 4–7–8 breathing; it can be used here and repeated over the course of a minute. It should be about three cycles.
- Now you are ready to do your test.

Step 3: The BOLT

- Sit calmly and comfortably, with the timer within easy reach.
- Through your nose, take in a normal breath—normal in the sense of not shallow, not deep, not fast, not slow. Let that breath out, at the same speed, through your nose.
- Now close your nostrils with your thumb and index fingertip, blocking the airflow completely. Don't let any air out or any air into your airways or lungs.
- Start your timer.
  - ⬦ Keep going until you feel the urge to breathe or the first message from your body saying "Get air."

- ❖ Reminder: These message-sensations may include the impulse to swallow; the constriction of your airways; and/ or contractions of your breathing muscles in your abdomen, chest, or throat.
- As soon as you experience any of these symptoms, of needing to breathe, *stop your timer.*
  - ❖ That's your BOLT measurement.
  - ❖ Release your nostrils and inhale calmly through your nose.
  - ❖ Resume normal breathing.

Record the number of seconds plus the date and time as "BOLT."

Now let's move on to our other, likely more familiar, rates to assess: pulse and heart rate. With self-assessment, it is common to need to trial a few times to ensure you are getting an accurate number. Do each measurement (pulse and heart rate) three times and then divide by three to get one number for each. Those will be your data points for pulse and heart rate.

PULSE: Let's review your anatomy. Roll your hand over to open your palm to the sky. When you place a finger on your wrist at the base of your thumb and move your thumb up and down, you will feel a tendon that connects the arm muscles and your thumb. Place two fingers on that tendon and then move your fingers back so your fingers are on flesh, not tendon. That's where you should feel your pulse. If you don't, move your two fingers to the right or down a little, to where you can feel the pulse. Take a moment to settle on it and register the rhythm.

Take your pulse: While your timer measures fifteen seconds, count the number of heartbeats. Multiply that number by four and that's your heart rate in beats per minute.

(You can also take your pulse by setting two fingertips lightly on the artery on the side of your neck and following the same fifteen-second procedure.)

Record the number, date, and time as "pulse."

RESPIRATORY RATE: Often called breathing rate, your respiratory rate is the count of breaths a person takes per minute. One breath includes two actions: an inhale and the exhale that follows it. To measure your respiratory rate, start by relaxing in a chair or lying in bed.

Observe each time your chest or abdomen rises as you take in a breath. Count each time you see it rising. Time the number of rises over the course of sixty seconds.

Record your respiratory rate, the date, and the time; label it as "Resp Rate."

And what about heart rate variability? Tracking provides you with a sense of how long your body stays in a stressed state (fight or flight) versus a relaxed state (rest and digest). Tracking it, however, requires as digital tool.

- There are watches and rings that sense and track heart rate variability.
- If you use one of these wearables, gather the most recent data (this past week, unless there were significant life or health abnormalities).

Before you move on to the next section of the assessment, make sure you have all these data points downloaded into your folder dedicated to Switch Optimization data.

DIGESTION

When you rolled out your crust, it revealed key Switch-function insights about digestion and hydration. That data should already be in your folder. There are two experiments to conduct to collect the remaining a data:

- The Bristol Stool Form Scale: Since its inception, this tool has helped practitioners, researchers, and individuals evaluate digestion based on the appearance of one's stool. We will use it as a key piece in refining your digestive assessment.
- The Sniff Test: This is a test I made up several years back after noting a trend that patients did not self-report or even consider the scent of their breath, underarms, and flatulence (bowel gas). These are signals we want to tune in to.

Bristol Stool Form Scale

Look at your bowel movements over the course of a week (see the chart on page 342). Determine which of the following most accurately represents what you see (for visuals, head to the "Operator's Manual" for a reference image). Select the number that correlates to your poop (your stool) most of the time:

Type 1: Separate hard lumps, like nuts

Type 2: Sausage-shape but lumpy

Type 3: Like a sausage but with cracks on the surface

Type 4: Like a sausage or snake, smooth and soft

Type 5: Soft blob with clear-cut edges

Type 6: Fluffy pieces with ragged edges; a mushy stool

Type 7: Watery, no solid pieces

Record your results, the date, and the time; label as "Bristol."

The Sniff Test

Sniff yourself throughout the week *before* you cover up your smells with teeth brushing, deodorant, coffee, showers, mouthwash, mints, gum, and so forth. Specifically, smell your underarms, breath, and any gas (if you pass gas or just smell your poop when you move your bowels). If you can't trust yourself to be objective, get the assistance of someone who will be brutally honest. Jot down some notes about how you smell naturally. You may also want to make note of any choices or factors that you feel impact your smells that day. An example could be that you worked out the night before and went to bed without a shower; you drank alcohol or ate "delicious to you right now" buffalo wings; you tried mouth taping for the first time last night.

Record your observations, the date, and the time; label as "Sniff."

HYDRATION EXPERIMENT:
ARE YOU A HOSE OR A SPONGE?

If you did the Hose or Sponge? hydration experiment earlier (flip back to page 102), you will use that data. If you haven't done it yet, do it now. You will need at least one result for the upcoming quiz.

Record your results, the date, and the time; label as "Hydration."

CHAPTER 10

THE SWITCH-FUNCTION QUIZ

Now that you have your data, it should take about five minutes to answer the questions to secure the results to process for your Switch-Function assessment outcome. You can head to the QR code on page 178, fill out the answers, and have an email sent to you with your data and the date completed. Or you can go old school: Grab a pencil and paper, jot down your answers, and note the date—yes, you can do this here in the book leaving space for additional assessment data. If you're the data-driven type who really loves this sort of work, build your own spreadsheet on your computer, phone, or device, and enter and crunch the data yourself.

Here are the questions.

SECTION A: HOW DO YOU FEEL AROUND EATING?

What does your eating, hunger, and appetite tell you about your current Switch function? Choose the letter that corresponds to the answer that more/most often describes your experience.

If you're torn because two statements that feel right, you can choose both.

When do you feel full?

a. Most of the time, after eating I don't feel full.

b. I usually only feel full a while after I've finished eating.

c. While I'm eating, I get to the point I feel full but often keep eating or drinking, especially if I want (or need) something with a different flavor (e.g., salty, sweet, bitter, etc.) or I am eating or drinking something to change my mood, to address stress, or to help me relax.

d. As I eat, I start to feel full and it helps me know to stop.

When do you think about food?

a. Almost all the time. If I see food, smell it, or hear someone talk about it, or if I am thinking about how I will schedule my day, food is on my mind.

b. Often, but not always, I am thinking about food.

c. When it's getting to be time to eat but also when I am tired or I'm drinking alcohol, using some form of cannabis, or I'm stressed, sad, or irritated.

d. Only when I'm hungry or when it's time for me to stop to have a meal or snack, so I think about what I want to or should have.

After you eat, how do you and your body generally respond?

a. With emotions, such as being upset with myself about

what I ate, or wishing I had eaten less, or being proud to have eaten the "right amount."

b. Feeling overfull, bloated, tired, or the opposite of "energized."

c. Feeling not really aware of how I feel or how my body feels (because I'm too busy, too stressed, too something . . .).

d. Mostly feeling satisfied and energized.

SECTION B: HOW ARE YOU BREATHING AND HOW IS YOUR BODY RESPONDING TO IT?

Yes, breathing needs to happen, but how it happens also matters. How you breathe is a key Switch performance indicator, and how you breathe has a big-time impact on your Switch function. Answer yes or no for each of the following:

Is your BOLT greater than 30 seconds?

Is your pulse less than 65 beats per minute?

Is your respiratory rate between 10–16 breaths per minute?

If you collected this one, is your heart rate variability (HRV) most days (unless you know you've been sick or had a bad night sleep) greater than 30 milliseconds?

SECTION C: HOW DO YOUR LABS SAY YOU ARE DOING?

Look at the data you gathered and answer yes or no for each of the following where you have a data point. The ones with an asterisk (*) are "core" and required; the others provide key insights and are worth getting if you are able. *Note:* For each lab, we are assessing a Switch-function range, and as such, the numbers here may be different from what your lab test results rate as "normal."

- High sensitive C reactive protein* (likely to appear as hsCRP): Is yours less than 1.0 mg/L?
- Hemoglobin A1C* (might appear as A1C or hbA1C): Is yours less than 5.4 percent or (36 mmol/mol)?
- Low-density lipoproteins* (LDL): Are yours less than 100 mg/dL?
- Triglycerides (TGs):* Are yours less than 100 mg/dL?
- Vitamin D 25 OH:* Is yours greater than 40 ng/mL?
- Fasting insulin: Is yours less than 10 µU/mL?
- Aspartate aminotransferase (AST): Is yours less than 25 U/L?
- Alanine aminotransferase (ALT): Is yours less than 25 U/L?
- Uric acid: Is yours less than 5.5 mg/dL?
- Glycated serum protein: Is yours less than 200 umol/L?
- Apolipoprotein B (ApoB): Is yours less than 80 mg/dL?

Earlier I advised you against relying on a ninety-day blood sugar average. So my inclusion of A1C as a core lab data point should give you pause. Ideally we would all have the other blood sugar assessment data points, including the additional labs above and CGM data. But the reality today is that for many of us, it's still difficult to get this other data, so we can use A1C—being aware of its limitations—as part of but not the sole data point for evaluating Switch function. If you have used/are using a continuous glucose monitor (CGM):

- Is your time in range (TIR) greater than 90 percent? (Note that the "range" should be 70–120; but if you are a person with diabetes, use 70–150, greater than 75 percent.)
- Is your average blood sugar less than or equal to 100? Or less than 140 for a person with diabetes.

SECTION D: HOW ARE YOUR DIGESTION AND HYDRATION?

What is your digestion and hydration data currently telling you about your Switch status? Answer the following with yes or no:

- Looking at your data from the "crust" that you gathered for this quiz, answer the following with respect to most days:
  - ❖ Are you constipated or do you have loose stools?
  - ❖ Do you rely on a "helper" (caffeine, supplements, etc.) to poop?
  - ❖ Do you experience reflux, gas, bloating, or burping?
  - ❖ Do you have acne, skin irritation, rosacea, hives?
  - ❖ Do you have ingredient/food/food group allergies, intolerances, avoidances because they challenge your digestion?
  - ❖ Do you have a current diagnosis of a digestive or autoimmune condition/disease?

Review the Bristol Stool Form Scale data:

- Is your stool type number anything other than 3 or 4?

GPS ALERT

HOW TO IDENTIFY YOUR RANGE

At this stage in the assessment, let's discuss how to address the categories for the ranges—male versus female. Weight composition, like many other metrics such as lab and wearable data, is currently evaluated in the context of sex and gender. As such, below you are asked to choose a range or a number that applies to your body today, and it specifies male or female. Address this the same way you do when addressing gender-focused questions in your health care. Whether it is the gender you were assigned at birth, the gender as which you identify today, or if you are nonbinary, make a selection. Remember our goal here is to establish a baseline and then track your progress with reassessment. So whatever you choose to use you want to be consistent with your selection.

Use your Sniff Test data for this question:

- Did you experience (or were you told) that you have foul-smelling gas, mouth breath, or underarm odor that week?

Use your Hose or Sponge? hydration experiment data for this question:

- Answer yes if a hose, no if a sponge:
  - If you had to pee *before* two hours (120 minutes), "hose."
  - If you did not have to pee for longer than three hours (180 minutes), "hose."
  - If you had an urge to pee within two to three hours (120–180 minutes), "sponge."

SECTION E: HOW IS YOUR WEIGHT COMPOSITION?

What did your weight composition reveal about your Switch function today? Only answer the questions for which you have current (last three months) data:

- Is your waist circumference
    ◇ For women, less than 35 inches (88 cm)?
    ◇ For men, less than 40 inches (102 cm)?
- Is your waist-to-hip ratio
    ◇ For men, less than 1.1?
    ◇ For women, less than 0.9?
- Skeletal muscle mass: Is yours at or above the percentage listed for your age, gender?

AGE	MALE	FEMALE
18–35	40 percent	31 percent
36–55	36 percent	29 percent
56+	31 percent	27 percent

- Visceral fat: Is yours less than 100 cm^2?
- Is your total body fat within the range for your age and gender? I've included the ranges below so you can see.

AGE	MALE	FEMALE
20–39	14–24 percent	21–32 percent
40–59	15–25 percent	23–33 percent
60+	16–26 percent	24–35 percent

CHAPTER 11

PROCESSING YOUR RESULTS

You now have the data you need to make two primary determinations about your Switch function:

PHASE 1—OVERALL STATUS:
OPTIMAL OR SUBOPTIMAL?

- If suboptimal, where on the suboptimal continuum

PHASE 2—SUBOPTIMAL STATUS

- Dysfunctional
- Delayed
- Suppressed

Now, read through the instructions for each section to your results. *Note:* Some sections deliver intel on only the first phase of the results; others reveal for both.

SECTION A: WHAT DO YOUR FEELINGS TELL US ABOUT YOUR CURRENT SWITCH FUNCTION? (PHASE 1 AND 2)

In this section, you will evaluate your data in two steps. First, look at the letters that correlate with each answer. This may not seem like "good data" to a scientist but it is money when it comes to understanding your Switch's current function. Why? Because a huge part of the Switch's job is to interpret and send signals based on and resulting in choices and your feelings about those choices.

SUBOPTIMAL OR OPTIMAL?

- Add up all of your As, Bs, and Cs. Then add up all your Ds. Is the total for your As, Bs, and Cs more than your total Ds?
  - ✧ Yes? You are on the part of the continuum described as suboptimal function.
  - ✧ No? Your starting Switch status is optimal function.
- For those with the suboptimal function result, let's get more details. Look at the letters you picked and select which of the below applies:
  - ✧ If you have two or more A answers, your current Switch-function status starts off as suboptimal-dysfunctional.
  - ✧ If you have two or more B answers *or* if you answered one of each (A, B, C), your current Switch-function status starts off as suboptimal-delayed.
  - ✧ If you have two or more C answers, your current Switch-function status starts off as suboptimal-suppressed.

Record your result as section A, noting the date.

SECTION B: HOW IS YOUR BREATHING AFFECTING YOUR SWITCH FUNCTION (AND VICE VERSA)? (PHASE 1 ONLY)

- Any no answers? If any of your results in this section are a no, your current Switch status is suboptimal. As you progress through Switch Optimization, let's see how your breathing improves your Switch function and how your Switch functioning better improves your BOLT, pulse, and respiratory rate.
- All yes answers? Then no part of this section suggests you have suboptimal Switch status. Record your answer to this section as optimal.

Record your status for section B and the date, and then move on to the next section.

SECTION C: WHAT DOES YOUR LAB DATA TELL US? (PHASES 1 AND 2)

Someday, just like they figured out how to turn Gila monster venom into a longer-lasting hormone, the white-coat people may figure out how to easily and effectively test our Switch functions (as they have done with other hormones). Even better, they'll come up with a gadget to monitor the Switch, the way a continuous glucose monitor monitors blood sugar. Until then, it's on us. Remember, do this section after you have gathered recent data for at least five of the core lab values (marked with an asterisk). *Note:* When processing these results as optimal or suboptimal, we are looking at the upper or lower end of a range. In your lab results, you are used to seeing a range defined as "normal." We are only scoring your labs here for their Switch-function value, not for your overall health. For that you talk with your practitioner.

- If you answered yes to *all* of the data points in this section, or at minimum if you answered yes to *all* the five core, there's good objective data that your Switch status is optimal based on blood work lab data. That's all you record for this section.
- If you answered any questions with no in this section (at minimum the five core), your Switch is expressing some degree of suboptimal Switch function.
- If your data includes only one no (the rest are yes), then your status is more likely suboptimal-delayed.
- If your data includes more than one no, then your Switch function is suboptimal-dysfunctional.

Record your status results for section C and the date, then move on to the next section.

SECTION D: WHAT IS YOUR CURRENT DIGESTION AND HYDRATION TELLING US? (PHASES 1 AND 2)

- If your answers are all no, then record optimal for this section.
- If you answered yes to any of the questions, you have a suboptimal Switch function today.
- If you answered yes to two or more questions, your suboptimal function is dysfunctional.
- If you answered yes to only one question, your answers here indicate suboptimal-delayed status.

Our digestion and hydration status is constantly shifting in response to external and internal stimuli. While you are collecting your results, they have shifted. Thus no result is *forever*. Whether your answers were all no, yes, or some combo, you don't get to skip future assessment and assume all good or bad.

Record your results for section D and the date, and continue to the next section.

SECTION E: WHAT IS YOUR WEIGHT COMPOSITION TELLING US? (PHASE 1)

- If you answered no to *any* of these questions, record suboptimal.
- If you answered yes to all these questions, then you either have an optimal Switch function or, based on your answers in the other sections, you have a type of suboptimal function but you are optimizing your body composition with your current choices, so you are more likely to have suboptimal-suppressed function. For now, record "optimal" for this section.

Record your status from section E and the date.

So now you have multiple indications of your Switch-function status, in five sections. You can aggregate these to determine your Switch-function status.

OPTIMAL VERSUS SUBOPTIMAL

- If your answer for each section was optimal, that is your current Switch-function status.
- If you answered suboptimal for *any* of the sections, your

current Switch-function status is suboptimal. Proceed to the next section to determine your phase 2 suboptimal status.

WHAT TYPE OF SUBOPTIMAL?

- Count up the number of answers you have for each type of suboptimal function—how many suppressed, how many delayed, how many dysfunctional. The one you have the most of is your current type of suboptimal Switch function.
- For clarity and how to handle a tie, review the following:
  - ❖ If you answered some with delayed and some with dysfunctional, your status is tied between the two. You can note this as dysfunctional (but closer to delayed).
  - ❖ If you answered some with delayed or dysfunctional *and* your section 1 answer is suppressed, your status is suboptimal-delayed (but closer to suppressed).

Phase 1 result:	Date:
Phase 2 result:	Date:

What do these specifics mean?

If you're dealing with a *suboptimal* Switch function, you are like almost all of us at some point. Your suboptimal Switch function—dysfunctional, delayed, or suppressed—will guide personalization in your plan with the goal of moving along the continuum toward optimal as you reassess your Switch function following experiments and efforts made in the coming weeks.

Here's a recap of the Switch statuses and what they indicate:

- Dysfunctional: You've got a breakdown or multiple breakdowns impacting Switch hormone production, deployment, and messaging quality.
- Delayed: While technically working, there is one or more challenge to Switch function resulting in suboptimal messaging, deployment, or receipt of signals.
- Suppressed: Switch function is poised to work properly from a function standpoint but your choices are interfering with operations.

If your results say your Switch function is optimal, you may choose to continue exploring this system, looking at the recommendations to better understand this system and your Switch. Today your Switch status does not indicate a need for optimization. So if you have other things on your to-do list, move on. But before you do, check out "Your (Re)Optimization," which includes two lists—one lays out signals that your Switch status may be or has changed, and the other includes Switch challengers that can trigger suboptimal function. Knowing these will give you a good sense of when to check in on your Switch by redoing the assessment. That said, I highly recommend quarterly Switch assessment—just like you change your closet for the season, see if you need to make changes to your choices to get or maintain optimal function.

That was so much fun—when do we get to do it again?!

Okay, that may *not* be how you're feeling, so I'll wrap up this section with quick notes on reassessment. Some parts of Switch Optimization may produce rapid results or feedback or both. For example, when you did the BOLT, you may have experienced a

GPS ALERT

DEGREE OF DIFFICULTY IS PERSONAL, TOO

Given the arrows between the four status results, there's a natural tendency to read the continuum as a difficulty scale. Those results describe types of suboptimal function but don't rank them by how hard they are to address.

Here's an analogy I don't love, as it talks about pounds and scales. I'll use it because by this point in the Switch story, I think we know how to adapt it. Two persons each set out to lose five pounds by a given date. For one, that's five pounds of ten total. For the other, that's five pounds of fifty total. The ease and pace for them is very different, while the near-term goal is the same.

You may make one or two adjustments based on the recommendations, experiments, insights, and support materials for personalization—and voilà, your Switch function moves along the continuum to delayed, suppressed, even to optimal. For another person whose function result is comparable, finding the one or two successful adjustments may take a fair amount more experimentation.

Scales of difficulty exist for lots of efforts. We find them on ski mountains and guides to mental games such as Rubik's Cube and Sudoku. Those are ratings for the problem—the slope or run, the configuration of the blocks, the numbers in the squares. Not for the solver. For each solver, degree of difficulty is personal, individual, and subjective.

We take on the challenge with the objective of getting there more easily, yes, and faster—but above all, better.

In the realm of suboptimal function, it can feel worse to have results that show dysfunction or delay and much better to have those that show suppressed. While these positions on the continuum indicate degrees of suboptimal function, the results do not reveal who has an easier path or whose road trip will be shorter or faster.

shift in how you felt; you now have a resource you can use regularly. Others, such as addressing the root causes of digestive or sleep disturbances, may take a few weeks or longer. And we know all too well that changing the scale—especially the one you are now using that looks at muscle, bone, and fat mass—may take longer than one better food choice, one week of exercising, and so forth. The key is to stick with it rather than giving in to old habits or giving up entirely. Reassessment can help us put together a realistic picture of how our Switch is adjusting to new choices and keep it while retaining old choices that are also better for our body.

As you navigate making new choices, consider doing a quarterly (or more frequent) reassessment of Switch function—the whole thing or one or more sections. Doing so reinforces accountability and may also show you that your efforts *are* paying off when you may be less likely to see or feel them.

PART V

HITTING THE BULL'S-EYE

YOUR BLUEPRINT IS YOUR BEST SHOT.

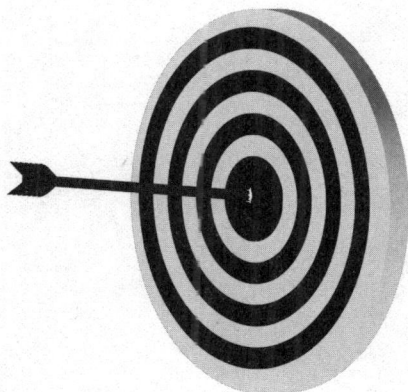

All the things you've done and discovered so far put the target in view and in range.

You may have even fired a few shots that landed on the board.

From here, we are going for the bull's-eye.

You've been building your force field—unlearning to create space, constructing your pizza—to identify where your body is today, how it feels about your current choices, and where to start experimenting with new choices. These phases built your foundation: giving your body more of what it needs and removing impediments. Now, as you choose your toppings, it's all about optimally resourcing your Switch and weight-health ecosystem to help them work optimally. You recall how I got nostalgic about the A-Team. You've got one. It's you. And you're your own John "Hannibal" Smith. Taking your best shot, you're bringing your plan together.

CHAPTER 12

TOPPINGS

"Ashley, does _____ work?" Via text, on phones, in patient sessions, (weirdly) across bathroom-stall dividers, and on airplanes—I get multiple Ask Ashleys each day; more if in a nonwork situation, I mention what I do or that I'm an RD. Ninety-six percent of the questions are not about pizza—literal or the one you've made, my metaphor for Switch Optimization. They are about supplements, the Shot, other medications, diets, and foods . . .

You've been wondering about those, too?

I get it.

The allure of the toppings is earned. They absolutely can make a pizza delicious to you in the moment. My digestion needed a few key toppings—probiotics, magnesium, glutamine—as supplements before my pizza could come together. All the sauce and cheese edits I tried wouldn't work until my pizza got the toppings it needed. In my case, like so many patients, those needed to come in the easiest-to-recognize, most readily available form for my irritated body. The same could be said for several of the patients' stories you've heard so far. The Shot or the GLP-1 activator supplement they added made it deliciously doable for them to build their pizza.

But ever had a delicious topping on a poorly assembled or not delicious pizza? Yeah, me too. Pass. On. That.

Too often, weight-health ecosystems get a lot of toppings thrown at them that are then relatively quickly discarded because they "don't work." These are cases of toppings applied to a suboptimal pizza foundation. Failure on the Shot? Toppings without a better pizza foundation. Failure on multiple diets resulting in a monogamous pizza relationship? Toppings without a better pizza foundation. Failure to optimize blood sugar despite taking berberine? Toppings without a better pizza foundation. And so on. You get it, right?

There is not one right order to Switch Optimization. But there is a better one.

Which is why we are here now—after assessment—for the toppings.

Selection of and experimentation with toppings will help you refine your plan into *your* blueprint for Switch Optimization. It's what makes the Switch Optimization plan *your best shot*.

So let's do it.

As you go through this section, be wary of the natural desire (due to decades of marketing) to read about an ingredient or protocol, believe it relates to you, and do it, expecting it will be "the solution" for your best pizza. Instead, the way you move through this section is to gather intel about different choices, home in on those that apply to you—based on your assessment—and choose one or maybe two to implement and then evaluate the impact. Was it doable? Did it improve your Switch function and weight-health ecosystem? If yes, then it is a keeper, at least for now and until reassessment shows you your body needs something different to become optimally resourced.

> *He deems it successful in the same way that a surgeon deems*
> *an operation successful when the patient leaves the operating*
> *theatre alive.*
>
> —Eric Ambler, *The Mask of Dimitrios*

It's common practice today to take a pill, look at one lab marker, and determine success. You've learned that we need to unlearn this approach in favor of optimally resourcing your body for optimal health span. This section is designed to help you identify where to invest your resources *right now*.

GPS ALERT

TAKE IT ONE AT A TIME

You are already making many better choices that we want to protect, and how you've learned to make and prioritize them will help you adopt new choices. A good rule of thumb, in our modern lives with ample to-do lists, is that the body and mind may be able to experiment with and evaluate at most one new thing a week, and it is nearly impossible to manage multiple experiments effectively at the same time. Go with one, assess, and either add it to your plan or pivot to a new experiment. Repeat.

With that preamble completed, it's time to build your force field.

Let's personalize your Switch Optimization.

I'm going to kick off with one last *not*—"Here's where we're not (going), *first*."

We're not going to begin with toppings with a focus on food, the Shot, or Shot-like supplements. Don't worry; we will get there and you will get what you need when you need it.

What matters more to your Switch and weight-health eco-system is nutrients. Nutrients provide the body the resources it needs to operate optimally . . . or not. The better course toward

---

GPS ALERT

BUT SHOULDN'T WE TEST *FIRST*?

A common scenario, now that direct-to-consumer testing is available, plays out for patients experiencing digestive complaints. They come to me having invested in tests for food allergies and intolerances, stressed about results implicating multiple ingredients. The kicker? The "triggers" ID'd are what they consume most often. Typically they've been instructed to reduce or eliminate these, and often have tried, or they're stuck between options like a deer in the headlights.

What did their tests really reveal? Digestive irritation. You put just about anything into an irritated system and it will respond by saying it is irritated. Anyone who's ever come home from a trying day at work and snapped at their kid/spouse/partner/parent/pet can testify to that. The test results gave us intel we already had. Their system being irritated meant that their food choices were irritating to it—because they asked an irritated system to do more work. This does not mean digestive testing is a bad tool. It's an invaluable tool when used with the right timing. The better order: (1) solid, multi-aspect assessment (as you've done), (2) get into toppings for digestion, as we're about to do, (3) invest in testing to further personalize a plan.

optimizing nutrient intake and utilization begins not at food but rather with your crust and (re)building it better with *digestion*.

My own story—with the belly, the worm, the probiotics— showed me that. My work with patients showed it to me again. And again. Hundreds and eventually thousands of times. For most of us struggling with weight health—the overwhelming majority of us—digestion is the primary issue. It's the right starting point. It's our base camp. Because our digestion is suboptimal, often wildly so, our bodies are unable to convert food into usable nutrients in the amounts and forms, and at the pace, our bodies need.

We don't win without optimal digestion—as seen on TV.

The first reality show hit the screen in 1948 with *Candid Camera*, but for most of us, *The Real World* introduced this genre. Reality TV has gone on to make celebrities out of "ordinary" people willing to share their lives for us all to voyeur. In the early 2000s, reality TV came for the weight-loss market. Why not? If the diet industry could make billions, wasn't there a slice of the pie for the entertainment industry to eat? Turns out yes, big time. For me, this version of edutainment seemed like a major teachable opportunity, a chance to bring my personalized nutrition message to millions nationally. I was already doing that with my magazine and news media interviews, so I dove in as the on-air dietitian for four reality shows. That almost none of my on-air content made it into this book tells you everything you need to know about whether reality TV shows are a better source of recommendations for your health goals.

BUT DOC, SHE'S NOT A "FAILURE"

I met Paula over twenty years ago. She was hitting her stride as an award-winning actress, singer, and a new mom. We met because we both had agreed to be on a reality show—she would join her peers because, for the first time in her life, she was struggling with her weight (post baby #2). She and her team would be a great storyline, it was a chance for her to get healthier, and there was industry pressure on her to look a certain way, so her team thought it was a win-win. I loved the idea of a storyline that could be about "body after baby" by optimizing from within. We both wanted to be a part of edutainment with a focus on the education part.

The reality was different.

You know how these shows work. The losers—the ones that lost weight on the scale at the big public weekly weigh-ins—were the winners each week. Unfortunately, where reality TV is like real life, the ones that got sent home were the non-losers—their weight stayed the same or, god forbid, they gained weight. The TV world, just like in the weight-loss programs and doctors' offices out here in the *actual* real world, has a name for them: the noncompliant. In many ways, those reality weight-loss shows contributed to major snags in the culture forming around weight health. But the biggest offense, in my professional opinion, was their validation of what I call the "Myth of the Noncompliant Patient." Namely, that there is a blessed, disciplined set of people who get a plan, follow it, lose numeric weight, and stay thin—a.k.a. "healthy"—forever. And then there are the slovenly, lazy rest of us, in the millions. A turning point for me came when the producers of one show were ready to kick Paula (a.k.a. "The Contestant") off the show for noncompliance. "She's obviously not trying," they had the medical doctor call to tell me in advance of filming for that week. "This will be an important story for us to tell," he went on. "Her mindset isn't there, as so often happens. We can use her to show others how they have to be ready to lose, that diets take real commitment. So we need you to show all the bad choices she is making, how she's not being compliant with the diet I created. I will then back you up to share the story of why she is a failure."

This was it—the moment of truth. I hesitated but not for long. I knew The Contestant needed her plan. And her failure was the doctor's plan. I hadn't gone toe to toe with doctors as often as I had to cave in on national TV.

"Uh, you don't have it right . . . ," I started, with my best A-Team effort. "Did you know she has IBS? The diet is making it worse. If she stops trying to follow it, and we add some digestive support, she will do great. We could show how . . ."

"Thanks, Ashley. I have to run. See you at the show." I was dismissed. And so was Paula. She actually took it better than me and said she was fine with however they wanted to characterize her—actresses have to develop thick skin early on. But she was so excited when I took a few extra minutes to ask her some questions and share my theory. It reminded me of the moment at the bar years before when that doctor waited patiently while I finished chanting "Kill the worm" and then asked, "Have you taken any antibiotics?" Still my patient today when she needs a tune-up, Paula has an incredible career and always credits her weight health—even publicly—to her digestive health.

And that's the reality.

DIGEST, PART 1

Tuning up your digestion is a priority for anyone pursuing Switch Optimization. This first part of the toppings focuses on digestive components, and how to identify and experiment with choices that may improve or resolve your core concerns.

INFLAMMATION

Like added sugar, protein, and fasting, inflammation is getting the attention it warrants. Our inflammatory response is a key indicator of our Switches' and ecosystems' operating capabilities. Cells, including those within the lining of the digestive tract, have localized inflammatory response mechanisms. Digestive-tract inflammation is a significant disruptor of Switch function. Recent discoveries about how the inflammatory response works give us insights for personalizing protocols to address inflammation for Switch Optimization.

SHOULD WE PREVENT INFLAMMATION— OR RESOLVE IT?

Prevention of chronic inflammation is a topic that has received a lot of attention in the last two decades with the popularization of the anti-inflammatory diet—most notably the Mediterranean diet—and its researched benefits. Much of this information is translated into the foundations of your Switch Optimization plan, which is designed to optimize the inflammatory response.

Newer to the inflammatory discussion is the resolution of inflammation—turning off the response—and what I refer to as the second phase of inflammation. Thanks to discoveries by Dr. Charles Serhan, a researcher in inflammation resolution, and others, as recent as the early part of this century, we now understand that inflammation, historically considered a passive process—it would just go away when not needed—is an active one, regulated by specific compounds. Resolution occurs when the body gets signals that the battle it has waged is won and it's time to call home the troops. Until those signals go out, inflammation rages on. Without active and timely delivery of that messaging, suboptimal ecosystem results occur. The long list includes chronic pain, impaired cardiovascular function, and autoimmune disease.

The compounds that drive inflammation resolution are called resolvins, which are specialized lipid pro-resolving mediators (known by the initials PRMs or SPMs). The body creates them with enzymatic activity on fatty acids (omega-3s EPA, DHA, and the omega-6 arachidonic acid). They get their name—pro-resolving mediators—because they promote resolution; they restore the peace by halting inflammation and its signals, initiating tissue repair, and cleaning up the mess that phase 1 of inflammation—"the battle"—created. This should be of big-time interest to anyone who experiences in-the-moment or chronic pain. As resolvins downregulate the inflammatory response, they turn off pain. And how you address pain will impact your resolvins. Unlike other anti-inflammatories, the presence of aspirin leads to the formation of specific aspirin-triggered-resolvins. You are signaling resolvins to go to work every time you use aspirin! The discovery of resolvins and this specific type helps us better understand the benefits in pain reduction and protective value historically ascribed to aspirin. More important, the discovery of resolvins—that our body can make them when optimally resourced and functioning and that we can get them from food and also take them in a supplemental form—is hugely important because other interventions to resolve pain challenge your Switch and ecosystem function.

In our food, we find resolvins in quality fatty fish, and we find the building blocks for resolvins wherever we get quality omega-3s and omega-6s. Thus the conversation about quality fats has many Switch-optimizing points—this being a big one. Yes, you can make resolvins from your intake of these fatty acids, but genetics, digestion, dietary choices, and health all impact their production. You can trigger resolvins with muscle contraction, so even though making muscle can initially be pro-inflammatory, building and using your muscles helps resolve inflammation. In personalizing Switch Optimization plans, I always recommend getting in the quality fats to build resolvins and efforts to make and retain muscle. I also often suggest supplementation with resolvins to resolve inappropriate inflammation.

MOTILITY

> *I like to move it, move it.*
> *You like to move it.*

Who doesn't like that Reel 2 Real song? Sad to say, many of our digestive tracts are not in that groove. They're not doing what we'd like, and they're not working as they're designed to. In most cases, it's not their fault. It's because of things we're doing . . . or not.

Are your bowels moving things in the wrong direction?

Not at all or not often enough?

Too quickly?

Or maybe making your life hell by alternating among all the above?

Oh, and do you ever experience stress—the bad or the good kind?

MOTILITY 1: TOO SLOW, TOO INFREQUENT, OR NOT MOVING AT ALL

Martha first came to see me because her bowels were not moving. Despite a super-active lifestyle, eating well, and exercising, she battled chronic constipation. In her early sixties, she was still deep into a high-profile career that demanded a lot of travel. It was stressful (in mostly exciting, good ways) and afforded her access to a lot of resources—trainers, spas for recovery, and now me. We resolved her slow-motility constipation immediately, correcting magnesium and calcium balance, optimizing fiber and hydration especially when traveling, and using breathing techniques to reduce the impact of life stressors. I'll share the specifics below. Martha and I stayed in touch, and I saw her occasionally in my practice. I loved hearing about her garden, her grandchildren who lived in far-flung places, and her travels to visit them. About a decade after our first meeting, a now-retired Martha came to me in desperation. She had high blood pressure for which she was prescribed two meds that she was refusing to take. To her, taking medications like this equated to surrendering to getting old and to reduced quality of life. She'd made a deal with her doctor: He'd lay off on the meds if I came up with an alternative. So, the pressure was on.

Her anxiety was palpable. (Not a win for blood pressure!) We reviewed her labs, food, supplements, and activity. No smoking gun. Then I asked, "Any more bouts of constipation?" She replied, "Funny you ask . . ." A few months earlier, she'd flown fourteen hours each way to visit a grandchild, and after the trip she had severe back pain and was constipated. She saw a chiropractor for the back pain, and when she mentioned the constipation, he suggested a fiber supplement—one he happened to sell. Aha! My inner A-Team sensed a plan that would not come together. Time to check out this fiber supplement. A search confirmed my hunch: It contained an herbal stimulant known to increase blood pressure. Nothing too exotic, just plain old-fashioned licorice root. We got Martha off the supplement then and there. We reoptimized with the Switch-friendly motility recommendations we'd worked up years before. Soon enough, all was regular. And she's never gone on blood pressure meds.

Back to Our Mission: Optimize Your Motility

If things are not moving easily, regularly, or at all, now is the time to remedy. Start by asking yourself the following questions to see what could be contributing to your suboptimal motility.

Have you started, stopped, or adjusted your intake of any of the following:

- Medications
- Supplements
- Nicotine
- Alcohol
- Caffeine or other stimulants
- Fiber ˙
- Water

Has your movement changed? Have you:

- Spent long hours sitting (for work, travel, etc.—remember Martha!)
- Added, increased, or stopped working out
- Been horizontal for significantly more than your regular sleep hours (injury, illness etc.)

Are any of these newish factors:

- Health concerns—a new infection, parasite, bacteria, diagnosis, and so forth
- More or different type of stressors
- Hormonal fluctuations—highs or lows; replacements (including the Shot)
- Difficulty sleeping
- Muscle cramps

BETTER TO *NOT* DO, FOR NOW: DELAYED MOTILITY

If you said yes to constipation, bloating, hard or pellet poops, or hemorrhoids, or if you currently need a "helper" to move your bowels, you are better off not initiating these common recommendations *for now*. If you are already doing one, or if your practitioner is recommending one, then the below is just gut-check guidance for consideration.

- **"Increase fiber"**: Sure, the whole world is screaming "More fiber!" While fiber is an essential ingredient for your weight health, as a blanket, all-purpose recommendation, more is not better. Ever. Especially not when your digestive tract is struggling to move and move things through optimally. Too much fiber at once risks aggravating bloat and pain. Fiber and what it picks up can get stuck in your belly with nowhere to go. It will increase stool size, making passing it painful and increasing risk of anal tears and bleeding. More fiber *can* mean greater variety of fibers or including some fiber at more of your pit stops. As we move through this section as well as using the four pillars of better nutrition, you can work toward an optimal amount and diversity of fiber for your body. It will depend on a variety of factors. Proceed cautiously when adding fibers: Diversify, optimize hydration, and *read how your body feels as your barometer*. And don't apply off-the-rack, one-size-fits-all rules.
- **"Increase calcium"—especially supplemental**: Yes, this bone-building mineral is key for many functions in the body. Motility is not one of them. Calcium works against motility as a counterbalance to magnesium, which relaxes muscles. When facing motility challenges, however, we want to look at total calcium intake to see if it is implicated. Many foods naturally contain calcium. Others, along with beverages, are calcium-fortified, and calcium is frequently added to a range of dietary supplements. So, it's easy to overconsume calcium—resulting in or exacerbating motility challenges.
- **"Cut out caffeine"**: This recommendation is often served up as a one-size fix for digestive issues ranging from motility

to reflux. We *will* assess caffeine in due course. Right now, it's key to note that your caffeine is likely *stimulating* motility. Stopping it would almost surely add to your motility woes. Once we get motility optimized, we'll explore the caffeine question.

- **"Take laxatives"**: Appealing as it may be to try *anything* to persuade your bowels to do their stuff, our bodies easily get dependent on laxatives, and they don't need much time to do it. Like many medications to help us sleep, laxatives perform at first and then they don't, unless we up the dosage. Laxatives don't give the same kind of relief, or the same degree, as we get by fixing our motility.
- **"Quit smoking"**: Yes, I'm actually saying this—with a caveat. Do *not* stop smoking without a plan in place to optimize motility. Smokers do themselves a huge favor by stopping. That said, nicotine is a motility promoter like caffeine (and often more potent). If you smoke and quitting is on your to-do list, by all means do it. Devise a plan to optimize motility first, using the recommendations below.

If any are a yes, here is one last instruction before you take an action:

- If the yes was started on the recommendation of your practitioner (not an expert or influencer who doesn't know you personally), make sure to inform them that you are experiencing motility challenges—especially if it is related to a medication or supplement—before adjusting the dose or deciding to stop.
- If you initiated the item(s) for which you note yes, hop into the ideas below on how you can modify your choices to reduce negative impact.

For any yes, let's work on improving your body's response by starting with what *not* to do!

Drum Roll, Please: The I Do Part

Now let's start with recommendations to consider initiating. As you move through this section, you'll almost certainly hit one or more "been there, tried that." This isn't your first "how to solve my motility" rodeo. I encourage you to review them anyway as I have caveats and experiments to guide optimization within each. That said, you aren't looking to do all of them—optimization is as much about passing on what doesn't work today as it is finding what's better. Our dual focus here is doability and hyper-personalization. The first goal—let's get your bowels moving. That may mean trialing several interventions to find solutions because it's likely that the root cause of your slow motility is actually root cause*s*.

Magnesium to Move Better

This mineral is a motility must-have because among its four hundred jobs, it hangs out in our cells to turn on rest-and-digest, especially when its alter ego, calcium, pops in, thereby creating tension (i.e., stress).

We used to get a lot more magnesium, and now, not so much. Today our soil contains less magnesium due to age and damage, and we process out most of the rest (from wheat in the field to enriched white flour, there is about an 80 percent loss of magnesium; cacao and cocoa when reduced to milk chocolate—versus dark—decreases magnesium, too). Low-carb dietary recommendations have further reduced our intake.

Not only is our intake down but we also consume more

challengers. Calcium, zinc, and iron are great nutrients but they happen to make it harder for magnesium to take center stage. Some efforts to improve magnesium intake by supplementing are not better because the form (such as oxide) is not well absorbed when compared to others. This means your optimization efforts will require looking at the quality of your foods, fortified foods, and supplements.

I see remarkable improvements in motility, sleep, blood sugar, and heart health when I help clients optimize their magnesium intake. So how much does *your* body need?

Our government's estimated average requirement (EAR) for "healthy adults" was updated in 1997—yes, back in the last century: 400 milligrams for men and 320 milligrams for women. Ah, sex-based recommendations for the non-win, as if women experience 20 percent less stress than men?! This recommendation was made using the average man's weight as 166 pounds and woman's weight at 133 pounds, assuming magnesium needs are dependent on total weight. My experience has shown that anyone with slowed motility will benefit from closer to 600 milligrams total in their day—and initially we may need to go higher.

How to optimize your magnesium? Let's start with better-quality food first. Consume at least four servings of foods (and beverages) rich in magnesium daily for a minimum of 400 milligrams (for examples, see the "Switch Optimization Nutrition Plan Recommendations"). For those with slow motility, I also recommend supplementing, starting with 200 milligrams in the evening. Your magnesium supplementation can also be split up during the day. For example, if you take a hydration supplement, you may get some there, and then you may take the remaining amount before bed.

In the "Operator's Manual" and online resources, you have access to a magnesium list for serving sizes, recipes, and a menu to help you experiment with optimizing your levels from food. You can also use a digital tracker to see how your motility responds as you increase your magnesium from current levels—both from foods and supplements, as indicated. So how do we optimize? Here's your plan:

1. Improve intake via food: Evaluate your current intake. If it's less than 400 milligrams, experiment to increase daily intake of magnesium from food sources to at least 400 milligrams. If doable, go for more within the context of your better nutrition choices. While too much magnesium is easily doable with supplementation, it's virtually impossible from food. One approach is to upgrade any current choices that are highly processed to ones that retain their natural magnesium content. Another is to experiment with adding magnesium-rich foods to your pit stops or increasing the portions of the ones you currently consume to reach your goals.

2. Supplement with better magnesium: Start with 200 milligrams of magnesium citrate or a blend that includes citrate with glycinate. Citrate is a form that has more of a "laxative" effect compared to glycinate—it draws water into the bowels, softens stools, and encourages elimination. There are notable differences between the laxative effect of magnesium citrate and most herbal or chemical laxatives. That and the multiple ecosystem benefits of optimal magnesium levels makes magnesium citrate (sometimes with glycinate) my preferred first motility-promoting intervention most of the time. If

you experience loose stools, then you have added too much supplemental magnesium.

a. Supplements taken orally come in several forms. With powders, chews, gummies, tablets, and capsules, you can determine what works better for your body and is doable. I typically recommend a powder form initially as it requires the least effort for the body to break down and use. Travel and life can make a powder less feasible, so the other options may be your better choice some or all of the time. In all cases, make sure to review the "other" ingredients, as we'll discuss later in this section, for any unintended consequences on digestion and your ecosystem.

b. While I typically do not recommend magnesium oxide, there is a form of ozonated magnesium oxide where ozone (O_3) is added to the oxide compound. Its purpose is not to increase magnesium for motility but rather to bring oxygen into the digestive tract to encourage loosening waste and elimination. I would consider it a short-term tool.

c. Beyond oral supplements, you can try magnesium salts, sprays, oils, and creams. An Epsom salt bath or one that has magnesium salts (read the labels) can be supportive for motility. Likewise, massaging an oil or a cream into your abdomen can also help to stimulate movement.

d. Phone a friend(s). Vitamin D can help your magnesium work more effectively. Optimal amounts of vitamin D will improve magnesium absorption. Review the labs you used in your assessment. If your vitamin D levels are currently less than 40 ng/mL, and especially if they are less than 20 ng/mL, experiment with increasing your intake of vitamin D via a supplement of 2000 IU–5000 IU, choosing one that includes vitamin K_2. As your vitamin D increases toward 60ng/mL, evaluate any impact on motility as well. This relationship—magnesium and vitamin D—is less perceptible but nonetheless important, and optimal vitamin D is a Switch Optimization must.

e. Vitamin C also plays a key role in motility, helping magnesium work better, too. Because it has all these Switch Optimization benefits and cautions with the wrong types of supplementation, it deserves its own moment in the sun—see the box below.

3. Identify how your intake of other nutrients is impacting your magnesium levels.

a. Calcium—your resources in the "Operator's Manual" and online will help you assess your total intake of calcium (foods and supplement) to ensure it is not more than 1,000 milligrams daily. If you've been advised otherwise, follow up with your practitioner before adjusting. If you need to be at higher calcium levels, you may need to increase your magnesium intake in response.

b. Iron and zinc compete with magnesium, especially in supplemental forms, so it may be better to consume them away from your magnesium supplements. (This is not necessary with food sources.)

c. Fiber, protein, and caffeine can also impair magnesium absorption. These, too, are typically more of a concern with timing of magnesium supplements than with food sources of magnesium.

VITAMIN C

Vitamin C aids in the absorption of magnesium *and* iron. This is a big motility win because if you need less supplemental iron to optimize your levels, you reduce a known motility challenger. And if you need less magnesium, it is easier to optimize those levels while also getting in your calcium and zinc. Ecosystem work makes everything work better—with less, not more.

HOW ELSE DOES VITAMIN C HELP YOUR SWITCH?

- Vitamin C is an ingredient in the formation of collagen, which supports the mucosal lining of the digestive tract, where the cells secrete the Switch hormones.

- Vitamin C promotes the production of enzymes the digestive system uses to break down food molecules— proteins, say—into usable parts, such as amino acids.

- Vitamin C helps to promote vagus nerve function.

When you optimize your vitamin C levels, you're not only improving motility but you're also getting a whole array of Switch-function wins!

HOW TO OPTIMIZE VITAMIN C

- Evaluate food intake to see if you are at least getting in the minimum level of vitamin C—the government recommended amount of about 100 milligrams. In the appendix and online, I share a few foods, servings, and their vitamin C amounts to give you ideas about how to meet your needs with foods and to reveal why you may currently not be meeting them (because of preferences or dietary intolerances). Like magnesium, the amount of vitamin C for optimal resourcing in your body is likely higher. A better amount for adults may range from 200 to 400 milligrams from food (or with supplementation) daily.

- Times when you may want to experiment with the 200–400 milligram range, and likely on the higher side, include:
  - If you have already tried improving motility with optimizing magnesium alone, and you have dry stools.
  - If you are also trying to improve your iron absorption.
  - Smokers and those routinely exposed to secondhand smoke are almost always insufficient or deficient in vitamin C. Likewise, those with digestive and thyroid diseases or dysfunction.

"But does optimizing magnesium really work, Ashley? What about MiraLAX instead?!"

Just last week I saw a thirty-year-old woman who was making big improvements in her health but was limited by one week of extreme cramps and constipation prior to her period. Timing being everything, it turned out that she "felt them starting today" as she anticipated her period in the next day or two. I had her get magnesium oil and two other digestive aids we will later discuss, and lather the oil onto her belly as soon as possible, using deep massage-like circular movements. Two days later I received this email: "I just wanted to let you know that I had the least period pain I've had in a really long time thanks to your suggestions!! Thank you SO much! The cramping was really minimal. I got the magnesium oil two nights ago and I think that really helped my constipation, too."

And then there was the physician who experienced bad gas and extreme constipation, who came to me on a twice-daily regime of Metamucil and MiraLAX. I assessed his dietary intake, finding his magnesium quite insufficient. He was a surgeon, so his stress levels were often elevated for sustained period of times. In addition to his Metamucil, he intentionally ate a lot of fiber, and he consumed yogurt and milk daily for calcium. We discussed the magnesium-calcium imbalance. I suggested he experiment with adding servings of magnesium foods and a magnesium citrate supplement. Within two weeks he was completely off his M & Ms (Metamucil and MiraLAX) and feeling great. The gas was gone, and he reported he could use one smaller notch on his belt. He was doing great—until he wasn't. A few months later, he, along with his wife, who insisted on coming with him, sheepishly reported the gas was back. Ever curious, I asked what, if any, changes occurred. "I stopped the magnesium. I got really worried my body was becoming dependent on it. So I went back to MiraLAX and then needed the Metamucil. We've always used MiraLAX with patients. I just feel more comfortable with it."

When it comes to my work, physicians are one of my favorite patient types, not because of compliance or their knowledge of the body but because if I can help a physician learn to optimize their ecosystem with better nutrition, that education has the potential to impact so many. So I went for education here. "Doc, our bodies *are* dependent on magnesium. For at least four hundred reactions. In suboptimal amounts, our systems may deprioritize relaxing the digestive tract to ensure other efforts, like relaxing the heart muscle to pump blood along at the desired pace. Your suboptimal intake is resulting in slowed motility." What our bodies do not rely on, but could become dependent on, is polyethylene glycol (PEG), a synthetic compound (a chemistry-lab product) derived from petroleum and the core active ingredient of MiraLAX. It is not a Switch Optimization tool.

When making choices for foods and supplements, quality and form matter. Giving your body what it recognizes and can use most efficiently should result in easier, better outcomes.

What else helps you to address slow motility?

- Optimize hydration. Optimal motility also demands that we optimize hydration—become a sponge. Use the recommendations you received earlier that include optimizing magnesium, one of the electrolytes. Use your hydration experiment to assess improvement.
- Optimize fiber. As noted in the "not to do first" list, fiber is a bit of a head-scratcher when it comes to slow motility. We know the body needs a diversity of types of fiber to help move waste through the digestive tract—along with other tasks such as feeding good bacteria. Yet increasing fiber in a slow-to-not-moving digestive tract can worsen the delays and failure to move. That's why I suggest first trialing the magnesium and vitamin C interventions, perhaps along with some of the others on this list. When considering fiber types to support motility, start with soluble fibers such as oats, flax, chia seeds, pumpkin seeds, berries, apples, carrots, kiwi, brussels sprouts, and green beans. For generations, prunes have been the go-to, yet today with continuous glucose monitors we see clearly how they spike blood sugar, which makes them a suboptimal choice. Experiment with psyllium husk as a supplement, but start with lower amounts than for someone who doesn't have motility issues.
- Optimize stress and physical movement. Elevated stress and moving less are strongly implicated in slowed motility. In the cheese section, we introduced the symbiotic nature of movement and stress. This shows up big time with motility. When stressed, calcium comes into the cells, and magnesium needs to be there to boot it out; the tension that endures restricts motility. Breathing, meditation, and less intense but perhaps

longer physical movement—especially time in nature—help you improve your body's internal motility by downregulating the stress response. We learned how to utilize these with the cheese. They would be good to experiment with here.

- Optimize being vertical. Our design uses gravity as a tool for optimal digestive function. When we stand upright, we encourage the appropriate direction of digestion. Evaluate your time spent horizontal or, at minimum, not vertical, especially noting how it may impact motility.

    ✧ Are you going from bed (sleep) to the couch or to a chair to sit without spending time on your feet?

    ✧ Do you sit at work, sit to get home, sit at home when you arrive and until you go to sleep, with minimal time spent standing?

    ✧ In addition to slowed motility, do you also struggle with reflux and bloating?

Yes? Your body is telling you that you aren't using gravity to your advantage.

Here are some gravity-related motility-enhancing experiments:

- Could you insert ten to fifteen minutes of walking or standing every hour or at least every three hours? After experimenting, note what gravity does for your motility efforts.

- Are you experiencing a period of slowed motility that coincides with your needing to spend more time reclining or lying down, such as the result of an injury or illness? Experiment with midsection movements to assess improvement.

- Midsection movement: Your digestive tract is largely a set of muscles taking up much of the middle of your body. The surface area of your intestines when unfolded would be longer than a whole football field. That's one big muscle! Muscles rely on contraction and relaxation to do their work. That's why we have focused on the roles that nutrients play and now will look at how you should manually support these efforts, too. Midsection movement—twisting and bending (or crunching, a more intense bend)—whether you do it in a way that engages the support of gravity or not, enables your digestive tract to move along what should not hang out any longer than is welcome. We should be regularly moving our midsections throughout the day.

- ◆ As noted above, there are times when this will be a specific challenge that arises, and you may have advanced intel for when this is going to happen—surgery, injury, travel, and the like. In all these cases, you can use midsection movement to help optimize motility. Remember to "blow it [candle] out" anytime you are twisting and bending. Exhale fully as you move into any of these stretches. (More on how to breathe coming right up.) Make sure you are near a restroom—these techniques can really work fast, and the stress of needing to find a bathroom is one you want to avoid.

- Twisting and bending can be extremely effective to improve motility and, while we're at it, address gas, bloating, and cramping.

- Incorporate hourly, if not more often (especially if you're experiencing digestive discomfort):
  - ◆ Twists, while seated in a chair or lying on your back with your knees tucked up. The key with both of these is that you don't move your shoulders but rather from your belly—side to side—while breathing.

- Rolling your belly on a large ball (basketball, volleyball, weighted gym ball, etc.).
- Touching your toes or touching the outside of the opposite foot with your hands. You can do this with straight or bent legs.
- Forward bending—standing or seated or laying your belly onto your bed or something that comes at about hip level.
- High-knee marching in place or tuck jumps.
- Crunches and mountain climbers.
- Weight machines weights.
- If you can't move your midsection on your own and you have engaged the help of someone (caregiver, masseuse, physical therapist), use these ideas for assisted twisting.

GPS ALERT

NOT ALL MOVEMENT HELPS MOTILITY

Some movements, desirable for other reasons, work against motility. The by-products of lifting heavy-to-you weights include building muscle, improving grip strength, improving mental health, and increasing fat-burning potential . . . but also impaired motility, because the body experiences lifting as a stressor and what follows stress is a challenge to motility. Similarly, the intensity of workouts will also produce a stress response. (If you monitor your heart rate, this is where it goes above "moderate"; if you zone train, this would be as you get into zones 3 and 4.) This stress will also challenge your motility. Thus using the movement techniques described above and the breathing exercises described below (along with addressing hydration) to induce recovery and reset the body into rest-and-digest mode will be important.

Remembering to exhale fully and invite the person assisting you to add magnesium oil if helpful to gain further relief.

MOTILITY 2: WHAT TO DO WHEN THINGS ARE MOVING TOO FAST

We all experience loose stools or diarrhea on occasion. It is tied to something we ate or drank, or how much, or to illness. We may momentarily adjust our nutrition and lifestyle choices or just let it run through us, and it resolves itself. The motility issue we're looking at here is different from that occasional experience. It occurs more often; it may even be chronic. When loose stools or diarrhea is *your* normal, the side effects are profoundly negative, even disabling. They include routine anxiety about your ability to go places and do things, and to have meals away from home; hypervigilance—a form of stress—about access to a bathroom; unwanted weight loss; hunger despite having eaten—with resulting weight gain; and effects of poor nutrient absorption or poor nutrient levels showing symptoms in hair, skin, nails, and fatigue. It quickly adds up to a general reduction in quality of life.

What might be triggering your going-too-fast motility?

If things are moving through you too quickly, now is the time to remedy. Start by asking yourself the following questions to see what could be contributing to your suboptimal motility.

Have you started, stopped, or adjusted your intake of any of the following:

- Medications
- Supplements
- Alcohol
- Caffeine or other stimulants
- Fiber

Are any of these newish factors:

- Travel with dietary changes and exposure to different water-types, environmental triggers, and pathogens
- Health concerns—a new infection, parasite, bacteria, diagnosis, and so forth
- Treatments and preventative procedures—colonoscopy, endoscopy, radiation, chemotherapy, surgeries, dental work (under anesthesia), and the like
- More or different types of stressors
- Hormonal fluctuations—highs or lows; replacements (including the Shot)

If any are a yes, here is one last instruction before you take an action:

- If the yes was started on the recommendation of your practitioner (not an expert or influencer who doesn't know you personally), make sure to inform them that you are experiencing motility challenges—especially if it is related to a medication or supplement—before adjusting the dose or deciding to stop.
- If you initiated the item(s) for which you note yes, hop into the ideas below on how you can modify your choices to reduce negative impact.

For any yes, let's work on improving your body's response by starting with what *not* to do!

Just like the universal "eat fiber" recommendation for folks with slow motility, here are the ones that you'll typically hear if you have fast motility. And, again, there is no cookie-cutter approach that works for everyone.

BETTER TO *NOT* DO, FOR NOW: FASTER MOTILITY

If you answered yes to diarrhea or loose and uncontrolled stools, here are the commonly heard recommendations I suggest you do *not* initiate first:

- Take something to stop your motility. Unless recommended and overseen by your practitioner, don't start trialing remedies to slow or stop your bowels. The body eliminates what it identifies as waste or otherwise counter to its health interest. You want to get to the root cause, not try using fiber, over-the-counter medications, or supplements to stop or slow your body's efforts to eliminate.
- Eliminate foods and nutrient groups. While a data-informed, time-specific "elimination diet" can be an effective tool, it needs to be personalized to ensure your body is replacing sources of nutrients with allowed foods to avoid creating or exacerbating insufficiencies and deficiencies.

THE CAN-AND-SHOULD LIST
TO STABILIZE MOTILITY

Here's what you *can* and *should* do to restore motility balance. The order below is my recommended approach.

Evaluate Stimulants

Things moving too quickly often result from or worsen when the body is overstimulated. Stimulants include caffeine, nicotine, herbal stimulants, and sugar.

Stimulant #1: Caffeine

Caffeine can be consumed from a variety of sources including coffee, teas, cacao/cocoa, sodas, herbs such as yerba mate, guarana, and others. By its very nature, caffeine encourages speeding up the pace of things, including moving your digestive tract. Caffeine stimulates two hormones—the Switch hormone CCK (!) and gastrin (stomach acid)—and increases the pace of intestinal contractions. You may love it for how it helps you focus and gives you energy—this can feel like a catch-22, especially if loose stools are currently zapping your energy! But its stimulating effect as a possible contributor for your motility woes needs evaluation. I write with deep sadness to share that my personal favorite food group, chocolate, known as cacao or cocoa, also provides some caffeine. It may be a factor here, just like it could fall under the next group (sweet stuff), too.

- Experiment with adjusting the timing, type, and amount of your caffeine intake to see if adjustments reduce or even resolve motility issues. In the resources, you have access to a caffeine evaluation that helps guide your experiment and includes recommendations.
- *Note:* Caffeine breakdown is genetic, so you may be someone who experiences a longer impact of caffeine, even

with seemingly little and less frequent consumption than others. You can do a nutrigenetic test to determine your sensitivity design, but you may know experientially: if you still feel caffeine's impact hours later or get anxiety or a racy heart with more consumption; or, conversely, if you can have caffeine and fall asleep a short time, even moments later.

Stimulant #2: Nicotine

If you smoke, vape, or chew tobacco, looser stools may be a side effect. This means that initiating efforts to quit nicotine should help resolve your motility issues and introduce the other health wins associated with quitting. So that we don't trade motility challenge types, as you consider quitting, review the slow motility comments on quitting smoking in the previous section.

Stimulant #3: Herbal Stimulants

Non-caffeine herbal stimulants can do many positive things for energy, focus, and stress. But they can also exacerbate or even cause motility issues. (Remember Martha!) Consumed as teas, supplements, and ingredients added to protein powders and fiber supplements, they are everywhere these days. Some are known as adaptogenic herbs or "adaptogens" and others may be less well known as stimulants. Here is an overview of names. Just because they are listed here doesn't mean they should be avoided or consumed, but their potential for acting as a stimulant should be factored into your personal experimentation: Panax ginseng, bacopa, rhodiola rosea, maca, cordyceps, eleuthero, schisandra, holy basil (tulsi), astragalus, licorice root.

Stimulant #4: Sugar and Sweeteners

The sweet stuff—naturally occurring in foods, added sugars, and nonnutritive sweeteners—is a go-to for mood and energy, but it is also a cautionary tale for digestive function, specifically fast motility. We've already discussed how excess added sugar contributes to dehydration—a likely side effect of chronic loose stools. Intense sweet cravings are another fast motility side effect. This makes sense when you consider that things moving too quickly mean that your body is not absorbing nutrients, so it feels under-resourced, fatigued (following nutrient losses), stressed, and sad. Your body will call out for sweets to get energy and emotional satisfaction. Microbial imbalances can also be a cause of sped-up motility. These imbalances result from sweetener choices and also contribute to sweet cravings. And finally, in an effort to manage symptoms, a person with fast motility may stick to certain foods and ingredients (e.g., the BRAT diet of bananas, rice, applesauce, and toast) or avoid all raw vegetables resulting in fiber, protein, and fat reductions—delivering sugar, and possibly further increasing sweet cravings.

Evaluate Alcohol Use

Alcohol is another contributor to fast motility. Alcohol is double trouble for Switch Optimization. It will dehydrate, and since a frequent side effect of too-fast motility is dehydration, this doubles the negative impact. Additionally, it speeds up digestive pacing, which is also problematic for a system moving things through too quickly.

- That alcohol limitations are dictated by gender is mostly unhelpful. In fairness, there is an observed gender difference. As a starting point, women tend to make less of the enzyme that breaks down alcohol than men. However, multiple factors ranging from genetics to health status to age and, of course, digestive function impact how your body responds to alcohol, so it is better to personally assess than to stick with the impersonal established limits for gender intake amounts and frequency.

- Experiment with avoiding alcohol for a few days, or at minimum reducing the amount and frequency, especially if you are wondering if a food or meal that you have in proximity to alcohol is triggering your motility woes. Also experiment with the amount and timing of your intake to see if your motility improves.

Optimize Fiber

Fiber comes in many forms. Often those experiencing loose stools avoid fiber-rich foods because they are told or infer that fiber gets things moving. Some fiber does. But other fibers, such as soluble fiber, can be a super way to *slow* motility and to experience some fullness that has eluded you.

- Experiment with adding in small amounts (about 1–3 grams) as a food or a supplemental source throughout the day to see if you notice improvements.

- Try oats (try overnight or "raw" oats), apples (to get apple pectin; baked may be better tolerated at first), brans (oat, rice, etc.), sweet potato (with the skin).

- While seeds such as flax and chia contain soluble fiber, they are also good sources of insoluble fiber, so they aren't my first choice for someone trying to slow down motility.
- *Tip:* Try adding 1–2 tablespoons of carob powder to applesauce (to make the carob easier to consume) as an effective and well-liked remedy for loose stools and diarrhea. This may seem like a weird contradiction because carob powder is mostly insoluble fiber. However, I learned about it years ago, and this tip really works! It's worth experimenting with to see if it's a win for you, too.

Experiment with Electrolytes

Experimenting with electrolytes may mean adjusting supplemental sources such as reducing the amount or shifting the type of magnesium while increasing potassium and watching calcium and sodium levels. Not only do looser stools and diarrhea increase the risk of dehydration but impaired balance of these minerals can also contribute to looser stools.

- If you skipped over the section on slowed motility, I suggest heading there specifically to look at the information on magnesium and calcium. Pay particular attention to added supplemental magnesium, which could be contributing to looser stools. Likewise, you may want to avoid supplementation with vitamin C.

Optimize Your Microbiome

An absence or imbalance of different strains of beneficial bacteria and yeast, as well as an excess of unhelpful bacteria, can be the root cause or a cause of impaired, too-fast motility. Medications (especially antibiotics), treatments (radiation, dental work, etc.), and even preventive testing (colonoscopy, endoscopy, etc.) can upset the balance of your good bacteria. Natural and artificial nonnutritive sweeteners alter your oral and gut microbiomes, too, and we will talk more about an optimal balance. Let's touch on three types of beneficial critters for your too-fast, too-furious bowels:

Saccharomyces boulardii is a beneficial yeast, not bacteria, that has long been used for acute diarrhea in children and, separately, for adults when traveling or other exposures trigger loose stools. It may also help with chronic sped-up motility, so adding it to your system to see if it helps you is a good experiment. Effective dose range: 2.5–5.5 billion CFU daily

Lactobacillus—reuteri, rhamnosus, casei, plantarum, and *acidophilus* are specific strains that may help resolve loose stools and diarrhea. Effective dose range: 1–10 billion CFU daily

Bifidobacteria—infantis, lactis, longum, bifidum as strains— may also be helpful. This probiotic was a game changer for me and my patients. I think of it like a quarterback in digestive, metabolic, and now weight-health-ecosystem optimization. Effective dose range: 1–10 billion CFU daily

In most cases, I recommend starting with food sources to optimize nutrient intake. Here is one of the more rare cases where I suggest experimentation with a quality supplemental source first and giving it two to six weeks to see if you notice significant (more than 51 percent) improvement before continuing to use it for further optimization. Working with a practitioner with knowledge of strain specificity and dosage is preferred for a personal recommendation. Some guidance includes:

- Look for genus, species, and strain specificity—there are a lot of different types of probiotics and they have different roles in the ecosystem. It is best practice to choose a probiotic—at least genus and species—based on evidence and, where possible, practitioner insights from their patient experience.

- Pay attention to the amount—there are large ranges in quantity. More is not always better.

- Are they alive when you take them and will they survive to your gut? This does not mean the probiotic requires refrigeration, but it does mean that the package or brand website should provide an explanation for how their probiotics stay alive (and for how long) in the package and how they will reach the desired destination alive. *Note:* There are some probiotics that are nonviable (not "alive") that are being used for therapeutic benefit. Discuss with your practitioner before taking a nonviable product to clarify if it is their recommendation.

- The other ingredients in the probiotic supplement should not include ones I have recommended you avoid. Proceed with caution: It is always a better idea to start lower and see how your body responds. Some patients experience an increase in symptoms at the onset of a probiotic. If that doesn't resolve in seventy-two hours, my professional opinion is that it is not the correct next step and the protocol should be reassessed.

Evaluate for Food and Ingredient Intolerances

Food and ingredient intolerances may be contributing to or are the primary cause of your accelerated motility. Your body is signaling it does not approve of or is unable to use an ingredient(s) you are consuming. For example, if you produce insufficient amounts of the enzyme to break down milk sugar, lactose intolerance may result in sped-up motility and loose stools. Similarly, for some, a sugar alcohol or modified fat, depending on quantity, could produce unwanted results. How to determine the cause or the exacerbators?

Gut-Health Microbiome Testing

Gut-health microbiome testing may be a useful tool. There are several kinds, and they can be pricey. These are better done with a practitioner who can guide you through interpreting the data. One value of testing is that rather than pinpoint specific ingredients as culprits, better testing can help assess intestinal permeability concerns. As we've discussed, when the whole system is irritated or suboptimally functioning, addressing the health of the lining and inflammation versus removal of a food or ingredient will produce foundational ecosystem wins. Allergy and intolerance testing, too, can be useful tools, but as mentioned earlier, I typically wait to do that until after other efforts are employed to stabilize (reduce irritation) gut health where possible.

Trial Elimination

A trial elimination is an effective experiment if done properly. An elimination may identify ingredient(s) to consider removing for a focused period of time. When done as a trial, an elimination is not intended as a long-term dietary approach but rather as a tool, an experiment for assessment. The results may illuminate one or more nutrients, foods, or even food groups to minimize or avoid while you are doing the work to optimize digestive health. Retrial may be used to determine if you are able to initiate consuming the original offenders and in what form and amount.

An elimination is comprehensive and time specific. This means 100 percent of a time period, typically twenty-eight days. Eliminate food, supplements, and all exposure to an ingredient (including skin care), and follow my guidance on how to replace nutrients, flavors, textures, and so forth, to make it more do-able and to not leave the body under-resourced. You may add supplements, too, but it is a short enough time that typically it does not require significant changes unless there is an underlying insufficiency or deficiency. Tracking symptoms before, during, and after is essential for the evaluation. Sometimes seemingly unrelated results occur and tracking helps connect the dots the body is trying to reveal. Do you reintroduce to challenge the elimination after? It depends. If results indicate significant improvement, then why reintroduce? Instead, build a plan around avoiding the known offender for at least ninety days and then reevaluate. Nothing is forever, unless a true allergy is identified. But if your symptoms are markedly improved, why not take the win?! If it is less clear or if the elimination feels truly unsustainable, then a specific, measured reintroduction of the ingredient— better quality, better quantity, etc.—may be your better next step. Does all of this sound complex or raise concerns about how you will build your nutrition plan without a key ingredient(s)? Good. I intended it that way, as it should really be undertaken in collaboration with a practitioner or coach with experience guiding individuals through eliminations.

Trial elimination is not used for a confirmed allergen. In most instances, a confirmed allergen means the need to avoid that ingredient.

Consider the Impact of Your Physical Movements

When digestion is moving too fast, you are likely to feel exhausted, under-resourced, and stressed about the idea of moving your body to build muscle and for cardiovascular health. This makes sense, and as such how to use movement to address too-fast digestive motility is personal and multifactorial. I suggest experimenting to see if shifting the balance of your movement can improve your digestive pacing. Higher-intensity exercise (meaning more stressful to your body) often negatively impacts your bowel function. For most people experiencing too-fast motility, slower-paced walks, yoga, strength training, biking, and other lower-intensity exercise can be a helpful way to reduce stress, build and maintain lean body mass, and support overall health.

Optimize Inflammation

Your ecosystem includes an inflammatory response to ensure that you get signals when things are worthy of alarm—fever, swelling, and so forth. This is what we call acute—in the moment—inflammation. We want the alerts it shares to occur on time and proportionate to the urgency of the issue. We do not want it to stay on longer. In this case, a motion detector is much better than a force field. Unfortunately, modern life presents many challengers that independently or collaboratively work to keep the inflammatory response turned on. Loose stools is an inflammatory response signal. It is time to decode it and optimize your inflammatory response for better outcomes.

Two nutrition components should be investigated. Note that we have discussed them along with others (alcohol, caffeine, allergies and intolerances, and excess intake of supplemental nutrients) already.

- Sugar: Excess added sugar contributes negatively to inflammation by encouraging the release of pro-inflammatory cytokines (proteins) that disrupt the body's intended, better efforts. As discussed, all aspects of your sugar intake should be evaluated.
- Fats: Similar to sugars, the four pillars—balance, quantity, timing, and quality—of your fat intake will impact inflammation. Excess saturated fat appears to drive pro-inflammatory cytokines just like sugar—thus the double whammy for those favorites that combine both. Additionally, insufficient or imbalanced intake of essential fats—omega-3s and omega-6s—will result in altered, unhealthy inflammatory responses. One strategy to optimize the fats that promote a healthy inflammatory response is to consider food and supplemental sources of resolvins.

MOTILITY 3: THE GOLDILOCKS ZONE

Some days motility goes too slow. Other days it's sped up. Does this define your digestive tract? This happens often for individuals experiencing any motility concerns. The list of triggers for either can be significant, and the interrelationship of the recommendations points out that some things that affect, even aid, one form of motility may be a trigger for the other. Hormonal shifts and travel come to mind, as they occur for all of us at one time or another. If you find yourself in an alternating Goldilocks zone, review the "Motility 1" and "Motility 2" sections and use the recommendations that apply to the ones you are experiencing *when* you experience them. Actions such as better breathing and nutrient optimization will help your motility all the time.

DIGESTION, PART 2: BREAKDOWN!

The question to address right now: Can your body use the protein it is *currently* getting? If not, it will send you signals that the protein may be stuck, festering as it sits in a lot of traffic, producing gas and body odor, triggering or exacerbating inflammation. Worried that might be you? That's why you are here. Before we add or adjust anything, let's make sure the body's equipped to use it.

Aim for Multiple Fuel Breakdowns Daily

Your body has to break down most of what you consume into usable components and forms. This breakdown is essential. It's part of what our weight-health ecosystem does for us. It's also one of the (many) ways in which contemporary habits throw off our weight-health ecosystem.

Here's how breaking down food into nutrients is designed to work:

When the body gets signals from the senses—smell, sight, taste, texture—saying that it has received food or is about to receive food, it goes to work on recognizing the nutrients coming in. Enter enzymes. The body starts enthusiastically secreting enzymes (made from proteins, one of so many reasons that protein matters in terms of quality and quantity). The first enzymes, amylase, work via our saliva. They target carbohydrates so that we can liberate fuel quickly, especially for the brain. Yes: Carbs, not coffee, are the brain's preferred fuel. Once food gets to the stomach, it gets hit with another set of enzymes from your stomach acid—pepsin (protease) for proteins and lipase for fats. By the time food arrives in the small intestine, thanks to digestive enzymes and fluids, what you took in resembles an undifferentiated ball

GPS ALERT

WHY MORE PROTEIN IS NOT BETTER

"But, Ashley, shouldn't I just eat more protein? That's what everyone is saying." They are *and* they (all) would be missing a big ecosystem insight. Protein is a big molecule whose parts—amino acids—have the potential to do a lot of really great stuff . . . or create big-time problems. Deciding to consume more without first knowing if what you are consuming is getting used optimally is not better. It's actually dangerous. When we increase intake of protein in a system that is not breaking it down and absorbing it optimally, we add gas to a fire. Too often people come to me as their bodies are exploding—hives, pain, belly weight, reflux, gas, allergies, intolerances, and so forth. Upon assessment, I learn that they've recently increased protein in an effort to achieve their health goal(s). There is a right order when it comes to better nutrition and that includes protein quantity and quality, as noted in the instructions and location of digestive optimization—that is, building your pizza crust. The body must be able to use what it receives. Otherwise, fat storage, inflammation, blood sugar imbalances, and other woes occur. Remember, *more* is a truly *im*personal, nonspecific word.

of slime, called chyme, that will go through the joint processes of nutrient absorption and tagging waste for elimination.

With that sequence of events happening:

- The body gets enzymes and fluids to break down protein into amino acids and to assemble them, along with others it can produce, into chains to create peptide (like your Switch) hormones.
- When breakdown happens at the right place and pace, the Switch gets its messaging and can do its reply-messaging effectively.
- The vagus nerve, your Switch transport system, gets required nutrients liberated from their food form—including omega-3 fatty acids, vitamins, and minerals—and can carry out its job well.

When breakdown isn't happening at all, on time, or at the right location, the follow-up responses don't happen or they happen suboptimally. This impairment contributes to suboptimal Switch function *and* impaired weight health. A few of the most common malfunctions are the creation of gas and bloating, belly fat (externally visible), skin irritation, liver fat, blood sugar imbalances, inappropriate chronic inflammation, and autoimmune responses.

Like a lot of my patients over the years, Fanny got my number on a referral after getting food allergy test results saying, as she read them, there was nothing on earth she could eat. She was losing it. Fanny said, "I'm allergic to *literally* everything I've been eating! Can you see me today? Like, now?" I suggested, first, we breathe. We did. Big belly breaths. Then I asked whether she'd gone in for the food allergy tests because she had digestive or skin issues. She responded with yes and yes. "Okay," I said, "let's meet tomorrow. For today, I suggest you eat what you normally do. You will continue to experience the same amount of digestive complaints you typically do and we can start to address it after our session." Her story is a variation on a classic I've heard too many times. Six months before, she'd taken a trip overseas and came back with diarrhea. Antibiotics, prescribed by her doctor, addressed the issue . . . for a few weeks. But then something else happened. A month on, bloating, gas, and she all but stopped pooping. Nothing she ate helped. "I looked third-trimester pregnant." First she cut out gluten, sugar, and her beloved chocolate. Then she removed most vegetables except for a few cooked greens. And on the guidance of something she read, she increased her protein—primarily quality chicken and fish. Things just got worse. Bumps on the arms. Red flush and heat on her skin. Bad body odor. More deodorant. Nothing worked, so the doctor referred her to an allergist for tests.

And now we were in a staring contest with the results. Chicken, check; fish, check; eggs, check; and a weird array of random ingredients including two of the vegetables she was still consuming.

"Of course I'm feeling bad. Everything I eat is poisonous for my body!" Noting that that was one interpretation, I offered another. "Your gut is irritated. I'd say really pissed off. And that means anything you put in is going to bother it because it's *already* bothered. The test really didn't tell you anything you didn't already know. Only now it's giving you more stress. More stress for an extremely stressed digestive system means impaired movement and suboptimal breakdowns. It's shutting down, spewing out gas and heat to get your attention." My hunch, based more on experience than on Fanny's tests, was that her main issue was not *what she was eating*. Rather, hers was a case of ineffective break-down. Fanny needed a digestive tune-up. Ideally she would have done one *after* antibiotics and *before* food allergy testing.

Why a digestive tune-up? When our digestion is challenged and not breaking things down efficiently and completely, it tells us it is unhappy. Typically it starts in a low but consistent voice, then it increases to a scream. Her initial reactions went unaddressed, so her body came back with fiery-hot skin, foul-smelling gas, and a distended belly that had her checking her cycle! Proteins such as chicken and fish require a lot of breaking down. When that doesn't happen, they move on only partially digested, which the body can't use—so it gets angry: "Why are you sending me these big chunks?!" In addition to reacting, it will try to get rid of the protein "chunk" however it can or try to work on it where it is. The results are not pretty or comfortable: Push it out faster. Push it out through your skin. Send smoke (gas) signals and smells out your backside and your underarms and your breath. It was not necessary for Fanny to swear off a long list of foods. In fact, by doing that she had worsened other symptoms (fatigue, muscle loss, joint pain, skin breakouts, hair loss). Instead, she needed to give her body the following:

- Digestive enzymes: Protein wasn't getting broken down. Likely fats and carbs weren't doing much better. At the start, we would add them supplementally and reduce some of the digestive workload using liquid and cooked nutrition to make things easier for her digestion.
- Resolvins: As her body wasn't doing a great job breaking down nutrients, it was likely that inflammation was not being

resolved. (I confirmed this with her hsCRP, which was 4.0; the goal is less than 1.0). While fatty fish was providing a baseline amount, we would start with supplementation and not rely on her fish intake or the body to make them from fat sources.

- Glutamine: This conditionally essential amino acid was very essential to repair her damaged digestive tract lining. You are going to learn about the digestive tract lining and the "glue" needed to repair any leaks in a moment.

- Magnesium, midsection movement, and deep breathing: Her motility needed support, *and* it would come with this three-pronged approach to help her body better manage stress.

- Good bacteria: The antibiotics had removed the offenders but also her supporters. Time to replace them with a probiotic blend.

And the results are in! It took three full months for Fanny to feel fully better—what we call 95 percent. (Remember, there is no perfect, no 100 percent, because an ecosystem is dynamic.) But Fanny let me know that after just two days she was pooping and after seven days her better nutrition choices were no longer giving her gas or bloating. The "pregnancy" look was gone. She was just Fanny, no longer freaking out.

Addressing suboptimal breakdown is a priority for any weight-health goal.

Fanny's story illustrates what can go wrong with something as simple as taking an antibiotic to kill harmful bacteria.

Most of us will experience something similar due to aging, infections, microbial overgrowth, eating habits (too fast, not chewing), insufficient nutrient intake, frequently choosing highly processed food products (especially those that contain ingredients that confuse or challenge the body), genetics, medications, treatments, and lifestyle choices. There are three common results:

1. The body doesn't produce enough of the enzymes to break down the amount of fuel received, or it may not produce any at all.
2. The body can't recognize what comes in, and it gets irritated and can't deploy the enzymes needed to break down the fuel it's receiving.
3. The fluids that escort the enzymes for food breakdown are not released in a timely manner or aren't released at all.

So, let's assess your breakdown capabilities.

Do your symptoms suggest you aren't having (digestive) breakdowns regularly?

☐ Gas and bloating, especially after meals
☐ Sightings of undigested food in your stool
☐ Reflux; belching
☐ Constipation
☐ Diarrhea; notably greasy stools
☐ Low energy, despite consuming nutrition
☐ Changes in appetite—loss of or unrelenting
☐ Skin irritation—acne, bumps, redness, burning
☐ Food allergies or intolerances, suspected or test results
☐ Taking medications and supplements to address any of the above

If you ticked any of those boxes, let's get started on optimizing your breakdown.

A TALE OF TWO FLUID WOES

STOMACH ACID	DIGESTIVE FLUIDS
Is it *really* too much acid? Nope. Too often, gas, bloating, reflux, and belching are self-treated or medically treated as elevated stomach acid by complaint, not a true diagnosis. Diagnosis is done with a capsule as a probe; sometimes people do it at home under the guidance of a practitioner with a baking soda at-home test (can be unreliable with false positives and negatives being treated incorrectly). Some things that lower stomach acid are H. pylori bacteria, infections, nutrient deficiencies and insufficiencies, thyroid issues, medications (antacids, proton pump inhibitors, the Shot), medical treatments, heavy alcohol use, autoimmune gastritis, stress, and gastric surgery. When a person takes medication or an over-the-counter remedy to suppress acid in a system, it lowers acid. If their system already has low acid, this will amplify insufficient levels of enzymes to break down food. *Caution:* Never cease a proton pump inhibitor by going cold turkey! It is stopping acid production; abruptly withdrawing it may feel like fire in your digestive system. It is better to discuss with your practitioner how to support your digestion as you wean off of it.	Here's a short list of what can interfere with digestive juices in *just* the small intestine: • Alcohol consumption • Over-the-counter nonsteroidal anti-inflammation drugs (aspirin, ibuprofen, etc.) • Prescription medications, including the Shot • Infections—bacterial, viral, and fungal • Bacterial or fungal overgrowth (SIBO, candida) • Parasites • Digestive diseases (inflammatory bowel diseases, celiac, etc.) • Impaired motility conditions and diseases (IBS, muscular dystrophy, scleroderma, etc.) • Any disease of producing organs (fatty liver, diabetes, removal of the gallbladder, etc.) • Excess acidity There's more—much more—that can throw off our digestive juices in the small intestine alone.

Let's start with something easy: the list of no-need-to-do-right-now.

BETTER TO HOLD OFF ON, INITIALLY: SUBOPTIMAL BREAKDOWN

If you said yes to gas, bloating, reflux, burping, diarrhea, acne, skin irritation, or food allergies or intolerances, or you are avoiding foods due to digestive irritation, I'd hold off on trying the following before proceeding further into the plan:

- **Slap on a stopgap:** Antacids, fiber, laxatives, marijuana, stimulants—these may provide in-the-moment relief but they don't help your digestive tract break down nutrients. And if not part of the solution, they can be part of the problem or create another one for your Switch and weight-health ecosystem.
- **Medicate for relief:** Similarly, oral and topical skin medications (antibiotics, tetracycline, Accutane, etc.), acid suppressors such as proton pump inhibitors, and remedies that slow or stop or speed up bowel movements are designed to override the body's current efforts to improve function. They don't fix the underlying issue—breaking down nutrients—and can make the problems worse or bring on others. They may be useful in conjunction with other efforts, not exclusively.
- **Treatments and dietary plans:** Colon hydrotherapy, cleanses, and other dietary plans constructed to avoid "irritating" food groups or ingredients—low histamine, FODMAP, gluten, lectins (this is not the same as avoiding a diagnosed allergy or replacing intolerances with those that are better for your body)—likewise may create relief and initiate some reduction in symptoms, but they do not fix what is broken. That's what we need to attempt to do now.

And now here are your *I do's*. These are the things to experiment with first, in the order presented:

Use Your Body to Stimulate a Better Breakdown Response

Improve Your Chewing

One of the reasons we have teeth is so we can start the breakdown of our food. Chewing, which is basic *physical* breakdown, prepares food for the chemical processes of digestion. It also triggers the release of saliva.

GPS ALERT

ON A GLP-1 AGONIST, CONSIDER THIS "SIDE EFFECT"

Addressing breakdown is essential to Switch Optimization. If you are on the Shot or evaluating it, there is an additional consideration: The GLP-1 Shot, by design, significantly delays gastric emptying. This will impair breakdown. If prior to the Shot you already had breakdown issues or are taking over-the-counter or prescription acid suppressors, suboptimal breakdown will get worse. While deemed a "side effect" of the Shot, it's not really a side effect when the Shot is working exactly as expected. It is treated with more acid suppressors. Increasing acid suppression exacerbates breakdown challenges. This becomes a vicious cycle that can often result in someone coming off the Shot due to unmanageable side effects, tolerating symptoms of suboptimal breakdown, or not losing fat *even on the Shot*. This cycle is avoidable, ideally with better digestive assessment before starting the Shot, and certainly while on it. On final consideration, as discussed in this section, increasing protein intake in a body that struggles to break down what it is already getting results in worsening symptoms of impaired breakdown. Yes, you want to meet your optimal protein levels but do so as your body can manage, which will improve as you optimize breakdown.

- When we underchew, it's like we're sending food to digestion with a note that reads, "Couldn't be bothered to break it up. Sorry not sorry."
- Optimize your senses: The body releases saliva in response to not only chewing but also tastes, smells, and even sights and temperatures. This response gives you some interesting and varied things to experiment with:
  - ✧ In the morning, try letting the aroma of essential oils, fresh fruits (especially citrus), tea, or coffee wake up your digestive system a few minutes before your first bites and sips. Yes, smells send those signals to your Switch that you are about to get something.
  - ✧ Pause before your meal, snack, or pit stop, or slow down when eating, so you're smelling and tasting your food more. This stimulates a better enzyme response than eating quickly.
  - ✧ Make your food look delicious. It doesn't need to win social media awards, but it should engage your visual senses in the whole "I am about to eat something delicious" game.

Use Kitchen Tools to Break Down Food Before It Gets to the Body

Food Prep

Kitchen tools and cutlery break down foods into more usable forms, reducing the work a stressed and overwhelmed digestive system may be having trouble getting to. Blenders, food processors, kitchen knives, whisks, ovens, and stoves all "kitchen process" foods so that they arrive on the plate or in the bowl more mouth ready. Forks, knives, and spoons do their part, too.

Better Liquid Nutrition
In the liquid form, foods are almost completely broken down into making them very easily absorbed. Soups, smoothies, broths, congees, purees, and juices demand less digestive energy, making them an excellent tool to reduce symptoms and to improve nutrient levels. The work gets done by the blender or food processor, or in a pot on the stove. The liquefied food moves more easily into and through the breakdown process, to vetting and absorption. This is helpful especially when a stressor—that tough conversation, brutal traffic, high-intensity workout—diverts attention from digestion or during a time when your body is working to heal. Liquid food intake can also be helpful when your body wants to finish its digestion sooner—such as near the end of the day—so it can move on to recovery and sleep.

- *Caution:* Not all liquid nutrition is created equal. It might be hard to eat 4 apples, but it's easy to drink 16 ounces of apple juice. It's hard to eat 3 cups of leafy greens but easy to drink it. We can get more key nutrients this way, and we can overdo and overwhelm, too—there's a lot of sugar in that quart of apple juice. Depending on how we process liquid nutrition, we lose some nutrients (e.g., making juice, we tend to lose fiber) while making others more available (e.g., cooking tomatoes with some olive oil) or more digestion friendly (pureeing cooked broccoli with some spices reduces the vegetable's gas-generating tendency). So, let liquid nutrition work for you as a tool—go for that smoothie after a workout or tuck into a stew as you end the day—using the better nutrition pillars to guide your experiments.

Eat Foods That Have Natural Enzymes to Help Break Down Carbs, Fats, and Proteins

These include:

- Amylase—banana (the less ripe–more green, the more enzymes), mango, raw organic honey (pasteurization and heating will destroy the enzyme), miso (koji), kimchi
- Protease—papaya, pineapple, kiwi, miso (koji), raw honey, kefir, ginger, kimchi
- Lipase—avocado, miso (koji), kefir, kimchi
- Lactase—miso, kefir

Hire a Personal Assistant: Supplemental Digestive Enzymes

Would your life be better with a personal assistant? I can't answer that. But if you and your digestion are not breaking down food efficiently and effectively, and you're experiencing any of the symptoms noted above or in the assessment, you may benefit from what I refer to as "personal assistants for the digestive tract": digestive enzyme supplements. Like when we are little and someone cuts our food into better sized bites, or if we were to have someone inside us continuing to chew our food after we've swallowed it, digestive enzyme supplements—to be distinguished from proteolytic enzyme supplements—add to any digestive enzymes your body produces and deploys. Because you can time their consumption (at the start of your pit stop), they have a better assured delivery timing. The result? Better physical breakdown of food. These supplements can be extra helpful during travel, holidays, or other times you are eating foods you don't typically consume, and during periods of higher stress (including good stress, such as after an intense workout).

- Experiment with taking a quality source at the start of pit stops. They are assisting digestion, so they need to be taken in proximity to food. Use as a rule: one serving with smaller pit stops; one to two servings with more complicated or larger pit stops. See if you notice an improvement. These can be real helpers at times when your body is stressed, when you're eating a type or amount of food you don't normally consume, or when travel slows things down digestively.

Make It Even Easier—Consume Foods Already Broken Down into Ingredients

There are some instances where your better choice will be a food already broken down into its absorbable form: isolates. Isolates may be partially or fully broken down. For example, a whey protein isolate versus glutamine, an amino acid. Others include fatty acids, vitamins, minerals, fibers, and antioxidants. The value of this form is the body has the least breakdown work to do. However, broken down doesn't mean it will be absorbed. As we will see next, that requires another aspect of optimal digestive function.

It's also possible to bypass digestion altogether, getting nutrients intravenously or via injection. That's reserved for specific circumstances, in consultation with your practitioner.

Reduce and Remove Irritants That Impede or Impair Breakdown

Alcohol—its impact on breakdown is another reason that elimination of or significant reduction of alcohol is on the Switch Optimization plan. Alcohol first affects the composition of saliva, including the enzymes it contains. Alcohol relaxes the lower esophageal sphincter, increasing the likelihood of reflux in the esophagus—specifically acid reflux, as there's now more acid. After that, about one-fifth of alcohol consumed is absorbed by the gastric lining, the rest via the lining of the small intestine. As it irritates both, it impairs their functions such as enzyme and fluid secretions, the timeliness and completeness of which is essential for optimal breakdown.

Assess What Can Produce Dry Mouth

Dry mouth reduces saliva production, resulting in less enzyme availability for initial breakdown. It is caused by insufficient water intake, allergies, coffee, smoking, open-mouth breathing, medications, impaired dental health, anxiety, and so forth.

Identify and Address Intolerances and Allergies

These can occur from insufficiency or outright deficiency of enzymes—such as lactose intolerance, in which the lactase enzyme is not available to break down foods (dairy) that contain it. Alcohol is broken down via a specific enzyme: alcohol dehydrogenase. Women produce less of this enzyme, and aging adults see a reduction, both leading to scenarios in which undigested alcohol moves through the body and more significantly irritates the parts that expect it to arrive broken down.

Optimize the Nutrition Pillars— Specifically Quantity and Quality

Taking in too much at one time or taking in ingredients that the body doesn't easily recognize will challenge the pace at which the body can break down what it receives. Experiment with optimizing these by following the recommendations and personalizing based on the insights you uncover in the next section.

Optimize Proteins

Proteins are the key ingredients in digestive enzymes the body deploys throughout the digestive tract for breakdown. You want to ensure you get optimal amounts and types of proteins.

Break Down Better, Better Health Results

Recently I worked with a sixty-year-old man, Monty, before and after his thyroid surgery. Part of our objective was to optimize his weight health. Before surgery we wanted to add lean body mass (or at least protect his existing amount). Immediately after, we would work on his healing, then enact a plan to improve weight composition. He was thin but insufficient in lean body mass, and he had pronounced bone loss. In the weeks before surgery, we had adjusted his intake to achieve better nutrient balance; protein and fats replaced a greater reliance on carbohydrates at meals. Post surgery, he had a harder time getting in food. And his motility slowed; he experienced bloating, so he didn't really want to eat at all. His albumin was low, so his doctor advised eating more protein. Aiming to be compliant, he tried but threw up the eggs and the chicken. Then, even when he was able to keep them down, he reported feeling so backed up and heavy he couldn't get a good night sleep. Not better.

When we talked, I asked him about his mouth and saliva production (and of course I asked about his bowels, where he relayed his ongoing struggle to poop daily). He had post-procedure dry mouth and was still experiencing insufficient saliva production. *Aha!* We personalized his plan: Yes, it needed to be protein-forward, but if we accomplished it with better liquid nutrition, his body would have less digestive breakdown to do. This meant adding liquids to protein-rich foods like lentils (we added dal "water" to his dal), nondairy milk with protein powder or yogurt made into a smoothie, stews—meat simmered in broth—instead of grilled meats, and soft-boiled eggs rather than scrambled and hard-boiled. Before he started eating, I had him smell cut lemon or orange, which he found delicious. That triggered saliva production. We also put lemon and orange into his water, to sip on after his meal. Meals were followed by midsection movement exercises or a short walk. He was already supplementing with vitamin C and magnesium presurgery to improve his motility, and we increased them. His recovery improved exponentially. Which of these measures worked? All of them, collectively and collaboratively. We also saw, using a body-fat scale, that after the first week, he stopped losing lean body mass. After four weeks he had gained a pound of lean body mass.

DIGESTION, PART 3: ABSORPTION

When it comes to your Switch, the absorption stage of digestion is the epicenter. This is where the hormones get their signals for secretion and deploy out. You might ask, "Why not start here then, Ashley?" Because without better motility and breakdown, absorption doesn't get the newly broken-down resources that it's there to absorb.

Absorption is incorrectly taught as the place where the digestive tract absorbs nutrients, thus solely as a digestive system component. However, hydration is not the process of taking in water but rather the act of absorbing water—by way of optimal electrolyte balance, sensing, and signaling—and with it the nutrients it escorts. Thus, absorption is two-pronged: digestion and hydration.

Let's optimize yours.

Imagine heating a house with the windows wide open in the winter.

Pretty inefficient and wasteful, right?

How about pumping gas into your car before driving away with the fuel cap off?

At some point in our lives, we all make absorption missteps. They may be minor if we catch them quickly, or they may be making an internal mess.

Let's look at what substances, actions, and diagnoses are an expression of these insults and injury—the "triggers" for impaired absorption:

SUBSTANCES	DIAGNOSES AND RESPONSES
Antibiotics, oral and hormonal contraceptives, proton pump inhibitors, chemotherapy treatment, steroids, retinoids, nonsteroidal anti-inflammatories, antacids, opioids, antihistamines, antidepressants . . . just to name a few	Genetic variants; diseases and syndromes including but not limited to autoimmune such as Crohn's, ulcerative colitis, celiac, Hashimoto's, Graves', scleroderma, type 1 diabetes, rheumatoid arthritis, psoriasis/psoriatic arthritis, prediabetes, diabetes (gestational, type 2), polycystic ovarian syndrome (PCOS), metabolic syndrome, obesity, cancer, multiple sclerosis, epilepsy, gout, irritable bowel syndrome (IBS); nonalcoholic fatty liver disease (NAFLD)
Alcohol, gluten, nutrient insufficiency, nutrient excess, nonnutritive sweeteners, emulsifiers (carboxymethylcellulose and polysorbate-80), higher intakes of fructose and glucose, high-fat diet	Disorders, conditions, exposures (or symptoms): leaky gut/intestinal permeability, gas, bloating, constipation, reflux/GERD, gastritis, gastroparesis, high cholesterol, high blood pressure, migraines/frequent headaches, small intestinal bacterial overgrowth (SIBO, H. pylori/ulcer, diverticulosis/-itis, chronic fatigue, chronic pain (joint, back, overall body), fibromyalgia, premenstrual cramps or disorder, sleep apnea, low testosterone/estrogen/progesterone, adrenal insufficiency, elevated cortisol, insulin or leptin resistance, depression, anxiety, seasonal allergies, histamine intolerance, mold exposure, elevated heavy metals and toxins (glyphosate, etc.), food allergies or intolerances, elevated stress and inflammatory responses, disordered and eating disorders

It's easy to see how many of us have insulted our digestive tract linings, not just once or twice but often throughout our lives.

It is time to make it right not just because your digestive tract linings are where your Switch hormones are made but also because they are the part of you that delivers the body what you are working hard to ensure it has. You get no credit from your body for eating kale or wild salmon or hemp seeds or drinking water . . . until they get to your cells for use.

THE ABSORPTION I-DO RECOMMENDATIONS

Here are your absorption optimizing *I do's*, in the order I recommend.

Reduce or Avoid Alcohol

Reduce or avoid alcohol (also known as "Yep, that again"). If you drink alcohol—a little or a lot—your mucosal layer is insulted a little or a lot. Experiment with a trial separation. If it's conveniently January, this can mean "dry January," but you can do this in any month that ends in *y* or *r* for that matter. During this time, replace alcohol with water, herbs, and alkaline formers such as lemons and limes. But don't replace it with added or naturally occurring sugar or higher-fat choices or a combo of both. For example, a mocktail made from sparkling water with frozen berries to replace ice cubes is a win, while a mocktail made from soda (including diet), juice, or added syrups isn't. Choosing to skip a drink with your meal is a win, but choosing to replace your drink with cheesecake— unless it's a few bites of something "delicious to me right now" (higher than 7 on the scale of 1–10) or a night you intended to

---

GPS ALERT

DON'T STOP YOUR MEDS (!!)

I do *not* want you to stop anything listed in the table above if they have been prescribed as an intervention by your practitioner *unless* you both agree to that as a next step. In almost all instances, I recommend completing at least thirty days of your personalized recommendations (*your* pizza) while continuing any prescriptions or over-the-counter daily remedies, and then, as indicated, starting a plan to wean off of them. Notice the word *wean*, not *stop*. Anything that you are taking or applying is directing your body to do things that it wasn't previously doing or to override its prior efforts. Stopping that abruptly is problematic. Instead, following a personalized weaning plan should help it restart or initiate better choices on its own—and if not, you will get those signals and be able to pivot accordingly. Separately, if you are using any of the above "as needed," I invite you to evaluate during the first thirty days how the needs may shift—less often, less in quantity, not at all, more. These will be important signals for where you go next.

---

have dessert—isn't. You get it, right? Take some time off the sauce while giving your intestinal lining what it needs to absorb better.

Add Glutamine

Glutamine is a conditionally essential amino acid found in foods. The condition where the body may make insufficient amounts? Stress. Ah, so let's just call it "essential for thriving in the 2020s and beyond?!" This is a critical piece of your Switch Optimization plan. Yes, creatine, taurine, and leucine are super-important amino acids and deserve the attention they are getting when it comes to brain health and lean body mass, but even they rely on better absorption. So that's why glutamine is key. We can get it via foods, but if digestion isn't better—and that's why we are here—then you likely need supplementation.

- Colostrum and collagen—other ingredients offering protein and antioxidant support—are not glutamine, and vice versa. Colostrum can be a good addition, especially if there are microbial imbalances. It's the "first" milk from mammals (cows) that provides the immunoglobulins needed for their calves to develop and thrive. Getting in colostrum and glutamine together may improve your results exponentially, but I don't recommend choosing colostrum *over* glutamine. Collagen supports muscle fibers and skin. I like to make sure we get in the nutrients to promote its formation, or supplement with it, but it is not glutamine either—different amino acids.
- Quality and quantity matters. Make sure to follow the guidelines for quality selection, noting both active and "other" ingredients in supplementation. Similarly for food and beverage choices, what is done to these—including heating and processing techniques—will impact how effective they are.

Reduce Sugar and Sweeteners

Excess added sugar and some nonnutritive sweeteners weaken the digestive tract lining. Recall your Sweet Taste Bud Test (and

Reset) result: Are you choosing sweets because your cravings are OOC (out of control)? This is a sign of insult and injury.

Assess and Reduce Elevated Stress

We are here again. Not only does stress affect your motility but it, too, insults your mucosal layer. Even the good kind of "OMG, Oprah just gave us all a car" excitement can be a burden. The body's design is brilliant in that it diverts primary attention from all unnecessary efforts so we can focus on experiencing and addressing the source of stress, almost exclusively, then be done and move on. It is our modern living that challenges a well-designed stress response at the sacrifice of our digestive success.

- Experiment with the "stress check-ins," midsection movement, and better breathing described earlier in the digestion sections.

Optimize Your Microbiome

Optimizing your microbiome with better bacterium types and balances, and feeding them with polyphenols and fibers, will enable the production of short-chain fatty acids that promote the health of the intestinal lining. More on which types in a coming section.

Become a Sponge

Optimize hydration. It is literally on every to-do list. That's why it's a recommendation in your "crust," the foundation of the plan.

DETOXIFICATION—WHAT WE DON'T WANT TO ABSORB IS TOXINS

"Should I try this detox, Ashley?" I get this question on repeat from audiences, online, the media, and often anyone with my number on January 2 each year. My reply: "You should be on a detox every day that ends in y." Especially in modern day, our weight-health ecosystem's success depends on optimal detoxification. Here are a few key reasons. If a goal is to lose excess fat mass, and we are successful in shrinking fat cells, we release the toxins stored in them into our body. They need to be eliminated successfully. Detoxification is key to getting them ready for elimination.

In the brain, cells also need to routinely clear out toxins. They are so special they have their own systems and approach to detoxification—autophagy, glymphatic drainage, and sleep. That said, some nutrients that are good for full body detox are also better for optimizing the brain's efforts to avoid excess or harmful waste.

With the popularity of hormone replacement therapies—mostly a great thing, in my opinion—such as the sex hormones testosterone, estrogen, progesterone, it is all the more important that we ensure that our detoxification systems are optimized. If we don't convert used hormones, those messengers that came, saw, and did what they were supposed to do don't get broken down and eliminated. Instead, they recirculate, and that's dangerous. This applies to hormones the body makes and external hormones we use to replace and optimize the function of insufficient or absent ones.

Life is messy, both internally and externally. Some of the mess we can control or reduce, but most of it we can't and shouldn't.

That mess needs to be eliminated completely and efficiently or we increase risk of disease over time. That's the job of our detoxification system. We should get out into nature, but the reality is that nature introduces us to more environmental toxins (and it does the same for the foods we eat and drink). We can do the work to reduce exposure to environmental toxins with our choices, but unless we live in a bubble and get shipped things made in a bubble—a nonplastic opaque bubble, so no sun or air gets in—we are going to be exposed to toxins, which our body is designed to process.

Genetically, our bodies may take longer to clean up that mess. Many of us have genetic variants with the potential for slower- or faster-paced phases of detoxification; if they become out of sync, a buildup and inefficiency can arise. Our genetics are only about 30 percent of our health outcomes, so it's important we employ epigenetic choices (the other 70 percent) to optimize our detoxification system's phases with nutrients and activity.

There are two primary ways to optimize detoxification: reduce your load *and* optimally resource your system. Let's explore both now.

Reduce Your Load

Reduce your load, as is doable. Just like with stress, it is impossible to live a healthy life, or live life at all, without exposure to toxins. The actions the body takes internally produce toxins, thus its design with a detoxification system. We *can* make better-quality choices to reduce our exposures to toxins. Depending on our health status and goals, the need to be more dialed in on this is personal and specific to our life and health stages. But overall, it is worth it to look at your more frequent choices and see how you can reduce your load. It's exactly why I offer guidance in this book regarding quality and encourage becoming a qualitarian.

Optimally Resource Your Detoxification System

Giving your detoxification system what it needs to run better, more often, is much more within your control. The recommendations present choices to nurture your detoxification system via foods and supplements. Yes, I am a fan of some deep cleaning and focused "detox" as long as it delivers nutrients that support your detoxification system. Tools such as saunas, sweating (in general), dry brushing, and lymphatic massage can be effective ways to support the elimination of toxins, when partnered with an optimally resourced detoxification system able to convert toxins for effective elimination. Fasting and fasting-mimicking protocols can also support detoxification efforts through encouraging the process of autophagy and liver support.

MORE TOPPINGS—THE SHOT AND SHOT-LIKE INGREDIENTS

What are the tools that optimize and activate or replace your Switch?

Does anyone else find it interesting that ultimately it was a prehistoric animal's survival mechanism that is helping us address a modern epidemic of Switch dysfunction? The realization that the Gila monster has a force field, not a motion detector, that helped it survive from prehistoric times to today ushered in a new way of acting and thinking about the human body.

As the line for patents of different versions of Switch hormone replacements surpasses that of a launch for a new sneaker, another line has formed. This one is composed of those who have harnessed the power of plant and animal compounds into dietary supplements. If the Gila monster's venom has an adaptable version of human peptide hormones, couldn't other substances? Could there be a plant or animal compound that could help our Switches *sans* Shot? Or when coming off the Shot? Because of course a natural supplement is better than a shot, right?

NOT. SO. FAST.

There is to date—and I expect it will stay this way—no natural supplement that equates with the GLP-1 shot's Switch-hormone replacement in quantity and duration. But there are absolutely supplements—and foods—that optimize and activate our Switch hormones, supporting their processes of production and deployment and our ecosystem function overall.

The hormones our body makes itself are called *endogenous*. When it comes to resourcing our needs, the body's endogenous production efforts can't produce everything, so we rely on exogenous sources. This is why we eat—to give our body essential nutrients and, when conditions call for it, conditionally essential ones exogenously. And finally, recall that an ecosystem has numerous components—biotic, abiotic, and interactions—to drive optimal function. Ecosystems, to satisfy all their needs, first look internally for resources and then send signals that they need exogenous sources to augment any deficits in their endogenous efforts and levels.

This is where food, fortified foods, supplements, and

medications come into play. These exogenous tools help us optimally resource and optimize operations within our ecosystem.

When considering tools to help us experience improved and optimal Switch function, there are the three categories of "exogenous" Switch support: (1) hormone replacement, (2) activation, and (3) optimization. Let's look at each category independently and then compare them to ensure understanding of differences and similarities.

Switch-Hormone Replacement Therapy

The GLP-1 shot is a hormone replacement therapy or tool. Today, it only comes in synthetic versions that provide GLP-1 and GIP-like hormones. By the time this book is available and in the time that follows, I assume we will have versions that include synthetic forms of PYY, CCK, and other Switch-related hormones. As a replacement, it is designed to do what your own hormone does and, importantly, to be recognized by your body's receptors for those hormones. The current synthetic versions are dose dependent, but even the lowest doses provide more of the hormones than your body is likely making. (At higher doses, the difference is very significant.) The alteration in the peptide chain of this synthetic hormone, what makes it different from what our body produces, delivers its signature force field–like benefit—lasting much longer (24/7 versus two to five minutes). More in quantity and timing means it has a much greater chance of getting to the receptor sites throughout your brain, pancreas, and other organs. In doing so, it overrides your body's own hormones. With receptors satisfied and the ecosystem follow-up work initiated, why call out for more production?

Switch-Hormone Activators

In some of the patient stories, you've seen that their protocol included an ingredient, Amarasate®. It activated Switch hormones at a higher-than-"normal" level (where normal means optimal function, noting most of my patients' daily experiences were somewhere on the suboptimal part of the continuum). Not just GLP-1 but also PYY and perhaps CCK. This activation sustained for about four hours, when consumed as a supplement an hour before a pit stop. In the pursuit of discovering this plant compound, the New Zealand government evaluated many others, arriving only at four that may work. The other three presented challenges to ecosystem function (two of which were downright toxic if consumed), so they were eliminated and thus they moved the hops forward to human testing. Who cares? I do, and you should, too. There are a lot of supplemental ingredients that are on the market—with more to come—claiming similar or better benefits. Until the evidence is there—in human trials and in practitioner protocols who can vouch for specific patient outcomes—pass on them.

Switch-Hormone Optimizers

This is where the bulk of the foods and supplements promoting their GLP-1 or Switch benefits belong. They help your body make and deploy Switch hormones for successful reception by the receptors for each hormone. They do not enable these hormones to last longer or avoid breakdown for any period. Our task now is to help you vet them for your body today. This summary provides a few overarching points for consideration. Nutrients (and critters) involved in the production and secretion of Switch hormones belong in this group. Fibers that are eaten by strain-specific bacteria to produce short-chain fatty acids that prompt GLP-1 secretion belong here. Foods that provide amino acids that are part of the peptide chain for your Switch hormones belong here. And so do the enzymes that help break down proteins or ingredients that encourage amino acid absorption. What doesn't belong here? Highly processed food products with poor-quality ingredients that, even if they help deliver ingredients that promote GLP-1 secretion (like ones high in protein or an excellent source of fiber), contribute to suboptimal Switch-hormone function and ecosystem disturbances.

Lexicon and categories established, let's dive into what you really want to know: how to evaluate these therapies and resources for your own ecosystem. Modern day presents us with lots of exogenous options for resources of varying value in the form of foods, beverages, and therapies—medications and supplements. To answer if an exogenous source is of value for your ecosystem today, you need to decode your ecosystem's needs (the assessment) and have a full picture of how the resource works—the pros *and* considerations. (I don't use *cons* here as I believe a con really only exists if someone isn't fully informed about how a resource works and will work for their body; the result of not doing their homework or being lied to . . . a true con.) Following is a chart about the pros and considerations for foods, fortified foods, supplements, and medications—all exogenous options that can be used to restore and optimize your ecosystem by complementing what your body produces endogenously. For all of these, costs and access are factors for consideration, as well as what is noted within.

Navigating these options wouldn't be that challenging if it weren't for another "gift" of the modern era—marketing. Each

	FOODS	FORTIFIED FOODS
PROS	Provide ingredients for the body to process into usable resources; trigger senses—taste, sight, smell—to activate internal actions; the body should recognize easily if not chemically processed or altered into an unrecognizable form.	Add nutrient(s) not found in the processed or whole food form of the food; may provide a different option for you to get in a nutrient (think calcium-fortified orange juice).
	SUPPLEMENTS	**MEDICATIONS**
	Address or prevent nutrient gaps. May provide nutrients independently, or in specific combinations, without calories, in forms easier on your body, with anytime access to them.	Override or augment the body's efforts to correct where the body is not able to self-correct. More immediate responses will occur. Specific, measurable amounts (doses).

	FOODS	FORTIFIED FOODS
CONSIDERATIONS	Extent of processing. Is it "delicious to you right now"? Is it a balance of nutrients? Do any ingredients or amounts challenge your body today?	Does the body recognize the nutrient form fortified? Does the portion you eat deliver a better or too-high amount or create imbalance with nutrients not present? Is the food processed to remove naturally occurring nutrients or is it otherwise of poorer quality?
	SUPPLEMENTS	**MEDICATIONS**
	Create imbalances in body's total nutrition, easy to consume too much of a nutrient(s); get nutrients in forms the body doesn't recognize better; "other" ingredients may present challenges; it may take time to see results; doesn't provide calories/ energy/fuel—may not trigger body's signalers, may not be absorbed as well as food.	Side effects based on design (intentional) or those that occur with a specific person based on their body's whole health (unintentional). Replace the body's need to assess and respond, thereby making the body reliant on the medication. Creates an ongoing need for the medication to maintain results. It can stop working effectively, requiring dose increases or adjustments based on other health issues.

ingredient or product has its own publicist whose job is to make us feel like their ingredient, and theirs alone, is the singular immediate-and-forever solution for our body's health goals. The honey-trap draws existing customers, then new ones. It turns into a trend, making headlines and beelines.

The GLP-1 agonist is like almost no other trend we've seen. With the success and challenges of the Shot, marketers everywhere are looking to reposition their ingredients to even be considered in the vicinity of this celebrity to get a piece of the action—or to steal its action.

Unfortunately, most of what you *and* practitioners *and especially* the media hear about today is the product of marketing delivered by those with a specific interest, which is getting their target consumer to buy, use, or recommend their product. That's not better, and we need to break through it to determine what is actually better for you.

Sounds like you need your A-Team? I've got you.

Enter the AKA-Team

After becoming a dietitian, because I had a prior career in marketing and an early childhood dream to be advertising executive Angela Bower, I brought marketing into nutrition. My idea was encouraged by my friend and manager who saw the value of my hyper-specific thoroughly vetted product recommendations. Quite business savvy, she also saw that I was literally writing them down everywhere, not just on patient plans to take into stores but at bars and restaurants when asked about something socially. "Ash, your approval is like the Good Housekeeping seal. This is what people need to find to make better choices more easily." And because

GLP-1–OPTIMIZING FOODS	
PROS	The body needs nutrients to make, stimulate, and deploy these hormones, so foods that contain amino and fatty acids, fiber, electrolytes, water, carbohydrates, vitamins, probiotics, and antioxidants would be "friendly" for your body's GLP-1 efforts. You will be introduced to these in an upcoming section of this book.
CONSIDERATIONS/ CONS	"On Track," the new badge to "help people on Ozempic make better choices," is problematic for three reasons. The first is that it doesn't acknowledge that food should be GLP-1 friendly for the human body. So its tag should say "GLP-1RA Friendly" to connect with people on the medications. And to lump all people as having the same needs—fiber, protein, etc.—well, you already know, that isn't a win. Finally, the ingredients they highlight is an old marketing trick to get you to not look at what else is contained, or even the quality of those highlighted ingredients— not better.

GLP-1–OPTIMIZING SUPPLEMENTS	
PROS	Due to preference, access, cost, and health challenges, you may have gaps in your total nutrition—what your body needs to be optimally resourced. Supplements—specifically GLP-1 optimizers—can help ensure that your body is optimally resourced. These include supplemental versions of all the GLP-1–optimizing foods noted in the column to the left. Accurately marketed, these supplements can support GLP-1 and the weight-health ecosystem to run better.
CONSIDERATIONS/ CONS	They are not a medication; they do not override the body's own hormones—they support them. Specifically, any GLP-1 hormone produced by the body will still be deactivated within minutes. That doesn't change with the addition of any GLP-1–optimizing supplements. Their ability to help your body produce, deploy, and activate receptors to incite follow up action is limited to the body working how it is designed. Any claims beyond that are hogwash. There is no "nature's Ozempic" and there likely never will be.

GLP-1–ACTIVATING SUPPLEMENTS	
PROS	Is there a compound in nature like we discovered with stevia or red yeast rice that can do naturally what a medication can do? Sort of. When it comes to stevia (and other natural compounds, such as monk fruit), we know they can get taste buds to experience sweetness that is hundreds of times sweeter than fruit, not thousands (like the artificial ones). Red yeast rice inhibits enzymes the way that statins do, but to a lesser extent. Amarasate™® is a compound that works like a GLP-1 agonist, but to a much lesser extent. It lasts for about four hours when consumed before a nutrition pit stop. It is not absorbed in the digestive tract. It does not stay on for longer than four hours, so it should not disrupt sleep cycles or heart rate variability.
CONSIDERATIONS/ CONS	It likely doesn't prevent GLP-1 from being deactivated; it requires twice daily consumption. It can produce digestive complaints initially, and it can reduce signals for feeling hungry, so when used, or overused, it could interfere with someone meeting their nutrition needs.

GLP-1 HORMONE REPLACEMENT THERAPIES	
PROS	At thousands of times greater intensity than what the body could produce, and designed to *not* deactivate for about seven days (depending on agonist), these deliver GLP-1-like hormones to receptor sites all over the body—and the results are impressive: reduced appetite/hunger messaging, inflammatory and insulin responses, and more. They help a person experience more complete satisfaction of hormones connecting to their receptor sites and, as a result to make choices that previously were more challenging or even unrelatable.
	Maybe they actually fix things, too. Semaglutide is showing promise for repairing intestinal-lining damage and optimizing the microbiome. Will this be longer term? For all humans? In my practice and those of my colleagues, we like the shorter-term wins we are seeing with this one peptide compound.
	Insurance may cover them; weekly injection or a pill versus daily interventions.
CONSIDERATIONS/ CONS	Not all agonists work the same. A single one, like semaglutide, works differently than a dual—tirzepatide, trio, quintuple, etc. When overriding more than one hormone in the body, you need to consider the implications of what happens because of each specific hormone being amplified and overridden *but* also the impact of multiple systems being regulated by an external source with different design from your endogenous ones—exponentially different results.

GLP-1 HORMONE REPLACEMENT THERAPIES	
CONSIDERATIONS/ CONS	For all agonists, for about 7 days, your body gets signals all day *and night*— this "always on" can impair heart rate variability by lessening the experience of time in "rest and digest." This can lead to fatigue and poor energy—often mistakenly ascribed just to eating less or the wrong things. It delays gastric emptying, by design, day and night, increasing risk of digestive complaints. The "I've got what I need" signaling can make it harder for the person to remember to give the body what it actually needs. Without optimal resources the body goes into hyper-prioritization, leaving things like building or repairing muscle, bone, hair, skin, and nails for when it has leftover resources.

I am who I am, Ashley Koff Approved (AKA), the qualitarian stamp of my approval, was born. Through this nonprofit I evaluated products based on their ingredients *and* marketing messages. I was one of the first to do that; and over ten years, I reviewed over a million products. If you were AKA, you could use it on your label and in promotions. If you didn't earn the AKA stamp, nothing happened other than you weren't in our database and couldn't use the logo. Today you can still find my logo on products and websites, but I stopped working on the database as hundreds of third-party–"approved" logos and certifications—including those of practitioners and influencers—flooded the market. It wasn't as useful as it had once been.

If I was still doing an AKA for foods, supplements, and

medications for your Switch, here's what I would consider and steer clear from.

Note: I am using *GLP-1* here because this is the way they are marketed, *but* you should mentally replace that term with *Switch* because as you will see these do more than address one peptide hormone in your body.

What does this mean for you?

No matter what tools you use—foods, fortified foods, supplements, or medications—you need to evaluate your current health and choices against the pros and considerations *before*, *during*, and, if stopping, *after* any that you choose to use.

"Ashley, I hate nutrition," Jane told me. "I've beaten myself up for decades dieting and I'm done. I am so happy with this medication, I'm never going to go off it." After eight months of semaglutide, she was meeting with me because her doctor was concerned about her bloodwork and weight composition (muscle loss), and wanted my help to prepare her to come off the medication (the doctor's goal). While Jane could appreciate some of these concerns, she was happy about her health, weight, and body for the first time in her life. She shared that if this doctor suggested she come off the Shot, she would find another or get the medication on her travels in other countries.

When we initially chatted, she shared how food was like an abusive relationship with an ex, and she was hell-bent on never going back. "I am not going to follow a nutrition plan," she said. "I want to eat what I like, when I want. It's working, and I am eating so little, it doesn't really matter." I said, "I hear you." And she exhaled, visibly; her shoulders came down two inches. "Here's the thing: Your body isn't getting what it needs to run optimally. But to be fair, I don't think it was getting it from your choices before, either."

"Yeah, I have noticed that my nails are awful, and I am losing more hair. I am hoping that my hormone therapy will address this."

"It may, but your body also needs more resources for your hormones, too. I don't need you to eat specific foods—I get that any discussion of that is off the table—but would you be willing to add supplements to get in those nutrients?"

She agreed. Over the course of the next three months, we worked out that about 70 percent compliance was her "better," and she could make that happen. Her labs, hair, skin, and nails all showed improvements. She was at first as categorically averse to strength training as she was to a nutrition plan. We found a happy medium there, too, especially after discussing the "flabbiness" under her arms that was making her self-conscious in sleeveless dresses. A trainer she could meet with virtually was a match, and within weeks, we saw results.

Can't I just microdose Ozempic to lose a few pounds?

First, let's get our lexicon sorted. A microdose is a microscopic dose—a fraction of what is used as a starting dose—of an agonist compound such as semaglutide or tirzepatide. Amount matters when it comes to your goals, both immediate and long term. Today, *microdose* is being used in place of *low dose*. At first, based on what I am about to share, I battled to "correct" the public lexicon. That's not a better use of my time. Just know that a microdose can mean a microscopic amount but today more often means lower than a starting dose of a GLP-1 agonist. To help answer this popular question, let's turn to the insights of two practitioners you should know, especially when it comes to optimizing hormones and weight health. Dr. Tyna Moore has pioneered the concept of using a microscopic amount of semaglutide for metabolic health—specifically *not* weight loss—wins. Dr. Moore was one of the first on my radar to have dug up the data showing GLP-1RAs' impact as an inflammatory and immune modulator, and paved the way for successful applications reported by other practitioners and patients on everything from autoimmune conditions to pain.

Another reason for the popularity of the microdose (a.k.a. lower dose) question stems from a concern about popularized side effects, including scary "Ozempic face." For several reasons—how the medication was made available (and protocols), desire for faster "results," shortened time frames of insurance coverage, getting someone to invest in an out-of-pocket package—many using these agonists (the Shot) start at a lower dose and rapidly increase to the maximum tolerated or available. I asked my colleague Dr. Tami Meraglia, who has worked with this peptide hormone as a tool for health optimization for many years, what she thinks of both approaches. She said, "This [use of agonists] isn't a weight-loss race. We can use lower doses, nutrition, and exercise to help people optimize weight health for lasting wins. We can even keep patients on lower doses for maintenance once initial weight-health goals are achieved for continued benefits, including the anti-inflammatory ones."

Do I have to stay on it (medication, supplement) forever?

Maybe. I don't have a crystal ball, but neither do the manufacturers or prescribers of these therapies. That's a problem. I have my own thinking about why this medication is scrutinized in this manner where most others are not. If you use a substance or redirect any ecosystem function, then going off it without addressing the preexisting suboptimal function is a recipe for a return to the prior results. Specifically, if you replace your body's own hormones, it follows that you would need your body's own hormones to be able to work *at least sufficiently* if you want to attempt to stop using the replacement. If you want to be successful, you would want your body's own hormones to now work optimally. That's what Switch Optimization offers. Whether and when to go off the replacement is best determined by your ability to continue to make the choices that you made with the replacement's support and whether those choices continue to optimally support Switch function and resource your ecosystem. It could prove doable forever, especially if the replacement enabled you to optimize your weight-health ecosystem. But remember my patient Jan from earlier in the book—like it did for her, life could throw new challenges at your Switch and disrupt that ecosystem. At that point, it may become necessary to restart use of the hormone replacement therapy as a tool. Or you may be successful with GLP-1–activation or GLP-1–optimization efforts. It depends. The same goes for GLP-1–activation and GLP-1–optimization supplements, and even food and lifestyle choices. When you stop them, you no longer have their support. So our job here is to teach you how to identify and navigate choices that help ensure you have the ongoing support your body needs.

What about using a supplement instead?

Meet Barb. She responded "ALL IN" to our research study on perimenopausal women who were interested in a natural protocol to optimize their weight health. The only consideration was that participants could not have been on an agonist to date. Barb described her past efforts as follows: "I am so good on my plan *until life happens*. I have dieted and been really successful, but more recently I can't seem to jump-start any weight loss. I don't want to use a GLP-1 agonist, but I am feeling really desperate." Prior to meeting, I reviewed her bloodwork and digestive evaluation. I got the picture. Shifting hormones were throwing her digestion off course. Her Switch was suboptimally delayed and inching toward dysfunctional. I put together a digestive hydration support protocol to address motility and absorption challenges, and I had her try out better nutrient-balanced pit stops, especially in the evening. We also added a tool, Amarasate®, to see if it would help her make better choices—even when life happened. Our digestion and hydration plan led to optimal weight composition within ten months.

GLP-1–activation supplementation can support someone while they work on repairing their Switch and weight-health ecosystem. They may continue to use it—especially as their body navigates perimenopause, which is a time (possibly a decade or longer) when hormonal fluctuations challenge digestion, hydration, and mental health. What Barb is doing now will help her enjoy better health for hopefully decades to come. Switch Optimization isn't about weight composition but rather weight-health-ecosystem optimization.

So what about other supplements?

Whether to add a supplement or medication is very much a personal matter. When it comes to Switch Optimization and restoring your weight-health ecosystem, a topping—supplements and therapies—may help make your pizza better for your body and deliciously doable for you right now.

Supplements fit into three buckets: they can help you address a nutrient gap, they can help you prevent one, and they can be a therapeutic tool in lieu of or in collaboration with a medication.

Ask yourself:

- Can I get in all the key nutrients in my plan, in optimal amounts, most days?
- Do I have evidence of one or more nutrient gaps in my nutrition plan?
- As I work to optimize my digestion and hydration, do they need operational support but not additional work (i.e., from food)?
- Am I taking a medication that partners better (reduces side effects, improves how it works) with one or more supplemental nutrients?

GLP-1–OPTIMIZING NUTRIENTS

Recall there are other spots in the book where I said "Don't be an Ashley," meaning you get the full benefit by reading it all. Here, read what you want and what applies to foods and supplements you are currently taking or considering. In this section, I do not distinguish food and supplement choices. That's intentional because it's personal. For example, you may love a broccoli arugula salad. Your digestion may not. As you read about the benefits of both, you may choose supplementation initially, and as your efforts to optimize digestion succeed, you can experiment with adding the food sources. Another example: Polyphenols are great for everyone. Your rainbow assessment—how many colors you get in daily—may help guide you on which ones to experiment with because adding them can help you fill out your rainbow. But if a polyphenol choice isn't affordable, delicious to you, or one that you can tolerate, skip it. This is how we personalize toppings.

Polyphenols

As part of your Switch Optimization plan, I encourage you to get in one or more servings of polyphenols from plants—these include blueberries, citrus, beets, plums, cherries, apples, strawberries, black currants, pomegranates, turmeric (curcumin), olives, extra-virgin olive oil, cocoa/cacao, coffee, and green and black teas. Some of these have direct Switch benefits, and all help optimally resource the weight-health ecosystem.

Bitter Compounds

These foods have so many benefits including what they do for your Switch. Leafy greens (kale, arugula, dandelions, radicchio, chicory), some citrus (grapefruit, bitter orange, lemons, limes—including their peels), bitter melon, and cruciferous vegetables (arugula, broccoli, cauliflower, brussels sprouts, cabbage) are going to be great for your Switch and ecosystem.

There are bitter taste bud receptors in your gastrointestinal tract (plus respiratory, reproductive, and nervous systems). When in receipt of bitter compounds, they stimulate the release of Switch hormones.[1]

Research shows the potential for some bitter polyphenols

---

1   Peyman Rezaie, Vida Bitarafan, Michael Horowitz, and Christine Feinle-Bisset, "Effects of Bitter Substances on GI Function, Energy Intake and Glycaemia—Do Preclinical Findings Translate to Outcomes in Humans?" *Nutrients* 13, no. 4 (April 2021):1317, https://doi.org/10.3390/nu13041317.

to reach the intestines (not absorbed earlier) and stimulate the "bitter" receptors in the L cells. It is exciting that several are favorites—coffee, dark chocolate/cacao, green tea, yerba mate, arugula. (Okay, arugula's a favorite of mine, not everyone!) Others are available in supplemental sources, such as Amarasate®, berberine, bitter melon, Himalayan Tartary buckwheat, and hops. How these polyphenols are treated in food and supplement production through preparation will impact their capabilities. Additionally, your genes may have variants for how well you sense bitter foods (the TAS2R38 gene). If you are less efficient due to a genetic variant, it may be harder for this to be a win for you.

Probiotics

An essential part of your body's Switch and weight-health ecosystem, thousands of different beneficial bacteria exist in your digestive tract, setting up their homes in different places to do important work. This is your microbiome. More recently, through the discovery of how novel strains such as *Akkermansia muciniphila*, *Clostridium butyricum*, and *Clostridium beijerinckii* work collaboratively with *Bifidobacteria* in the lining of the intestinal tract, it became clear that they play a major role in your Switch. As they chomp down on mucin, *Akkermansia*'s secret P9 protein tells the L cells to secrete GLP-1. *Clostridium butyricum* and *Bifidobacterium infantis* help ferment dietary fiber for the production of short-chain fatty acids (butyrate), which also stimulates L cells to secrete GLP-1. I've conducted research with patients where I added a supplement of these bacteria strains as part of their Switch Optimization plans and the results on their blood sugar, cravings, and better choices are impressive. In one study focused on optimizing blood sugar, my team got some very Switch-optimized results. Over thirty women tried a blood sugar optimization nutrition plan. To optimize digestion and weight health, I added a supplement: two different doses of a specific probiotic blend—*Akkermansia*, *Clostridium*, and *Bifidobacteria*—Glucose Control or Metabolic Daily, both from the brand Pendulum®. While there were many improvements in blood sugar, the participants who initially reported digestive complaints, challenges with cravings, and "food noise" saw improvements in these areas. What a full Switch-function-ecosystem optimization story!

Glutamine, Creatine, Leucine, and Taurine

Optimizing levels of glutamine, creatine, leucine, and taurine, which are amino acids and amino acid compounds, is essential for optimal Switch function, regardless of whether the amino acid is technically termed "essential" or not. As discussed, glutamine is a conditionally essential amino acid—the conditions of life today make glutamine a crucial ingredient in Switch Optimization. Beyond digestive tract muscles, glutamine also provides a source of energy for muscle tissue, helping to build and repair damaged lean body mass. And glutamine helps improve the secretion of GLP-1 and GIP, your Switch hormones.

As you learned in the absorption section, I often recommend supplemental glutamine, starting at about 5 grams (especially for those who avoid dairy and/or fish and seafood, primary sources of this amino acid).

Previously relegated to "the gym bros who want to get huge," creatine has thankfully undergone a necessary makeover. Its benefits read like the list of Switch-hormone functions: supports the building of lean body mass, brain health, hydration, and blood sugar optimization. Creatine is synthesized from methionine, arginine, and glycine, and two enzymes. That means any challenges—genetic inefficiencies, dietary intake, digestive capabilities—to optimal levels of these will lead to insufficient creatine. It is a weight-health-ecosystem superhero, so like glutamine, it often tops my list of supplemental interventions as I personalize a plan for optimal weight health.

Along with glutamine and creatine, two other aminos deserve attention—taurine and leucine. Taurine, like glutamine, is deemed conditionally essential, and leucine is an *essential* amino acid. Taurine is concentrated in key areas to promote focus and targeted effort—eyes, brain, heart, and muscles. It helps promote optimal hydration. And here's the cha-ching: The body uses it in times of elevated stress. Taurine supports immune function, inhibits inflammatory cytokines, and helps restore calm in the nervous system. Thus, it's one that I move from conditional to essential to optimize levels, given the stress we intentionally and unavoidably encounter. Yes, we can get it from foods, but we often come up short. I typically see it at insufficient levels on my patients' amino acid panels.

The same goes for leucine, which protects and supports muscle building. Most protein-rich foods sources—plant and animal—contain leucine. The dietary supplement hydroxymethylbutyrate (HMB) is often confused with leucine—the former being what the body breaks down leucine to become—and is better used in those doing more intense training or to prevent muscle breakdown when caloric needs aren't or can't be met.

So, which amino acid should you supplement with? That depends on your assessment and ideally additional data—doing an amino acid assessment of your total nutrition especially on training days and/or an amino acid lab test. Otherwise I am a fan of using a blend of all of these aminos because targeting one amino acid versus another can be harmful to Switch Optimization efforts, especially via supplementation, by throwing off beneficial balances.

Omegas

Speaking of beneficial balances, we've created big, fat messes in the nutrition world. First with advice to remove and replace whole fats entirely and then with modified versions, then to serve up isolated fatty acids, and now demonizing one set of fatty acids that the body deems essential. I am speaking of omega-6s.

Omega-6s are essential fatty acids, and we likely need as much if not slightly more of them than omega-3s. They include a variety, several of whose Switch benefits I have mentioned: GLA (the glamour fatty acid for hair, skin, nails, and hormones) and arachidonic acid for creating pro-resolving mediators (resolvins) to effectively turn off inflammation. Their precursor, linoleic acid, is also beneficial for skin and heart health.

With better nutrition, you've learned that quantity and quality matter. This is at the core of the issue with omega-6s and the specific form in the hot seat right now—seed oils. As food and beverage consumption evolved over the last century, the processing and resulting ingredient inventions resulted in a "standard American diet" that is excessive in omega-6s and that emphasizes poor quality forms. *That* is the issue to resolve for its negative impact on Switch function and ecosystem health. Anyone who says that omega-6s are inflammatory fats and should be avoided either doesn't understand how the inflammatory response works or is trying to oversimplify nutrition, typically as part of a marketing effort to sell a product or their influence. Buyer beware.

Case in point: seed oils. Circa the 1950s with the introduction of soy and canola oils, marketers with products to sell—namely, cheap blends of poor quality oils—brilliantly named "vegetable oils" to get us to feel better about consuming them . . . because of course we should eat our veggies. Choosing vegetable oil over saturated fat from sources such as butter became the mantra for decades, but notably heart-health statistics did not improve. These seed oils are not better for your Switch function and ecosystem. Eat your veggies and ditch the vegetable oil. Choose quality oils such as olive and avocado, and, yes, enjoy better-quality butter and ghee. Access, availability, and cost can pose challenges, so I recommend starting with upgrades to what you consume more often. Some seeds and their oils are a great way to deliver the body these valuable fatty acids.

Like amino acids, we need a better balance of fatty acids from quality sources. That begins with awareness that no food ingredient in nature has just one type of omega fatty acid. The intestinal lining has a receptor (GPR120) for two omega-3 fatty acids (EPA, DHA) that triggers GLP-1 secretion from the L cells. Additionally, they likely spur PYY production as they support production of short-chain fatty acids.

Short- and Long-Chain Fatty Acids

Another set of Switch-function-supporting fatty acids are known as short-chain fatty acids and by the names propionate, acetate, and butyrate. They are massively important for your Switch and ecosystem functions, and are produced in the gut when beneficial fibers digest dietary fibers. (Ding, ding, ding—the bells go off on why those are so important, too.) Running through what they do and where they do it gives us further insight into their direct Switch-function and ecosystem efforts.

- Propionate is used in the liver for glucose production and promotes the release of Switch hormones.
- Acetate is an ingredient in energy production throughout the body's organs.
- Butyrate promotes Switch-hormone release and is a key food for colon cells helping to maintain optimal function of the digestive tract lining.

There is an odd long-chain fatty acid, C15, getting a lot of attention. (Most fatty acids are even in length.) The body doesn't make it, but we can consume it in butter, some fish, and more recently supplements made from plants. There's warranted excitement about C15 as it helps address cardiometabolic issues and fatty liver, thereby earning consideration as part of your weight-health ecosystem.

Salt/Sodium

The optimal quantity and timing for you to consume sodium depends on several factors. We've discussed sodium in the context of hydration and taste. It is not an evil compound as the salt-free era tried to teach, but it does need to be optimized, personally. We do better when sodium and potassium are closer to a balance. For most of us, that means a reduction in added salt or salt-fortified food products and increasing our intake of potassium, primarily from foods. Look at your supplementation, too—all of it—to make sure you aren't exceeding your body's better sodium intake. For those that use sodium supplementation to optimize hydration, you may want to adjust your food choices accordingly.

Potassium

Potassium is one of the harder nutrients to achieve optimal levels from foods and beverages—even harder because of the nutrition marketing messages popular today. Most excellent sources of potassium also are significant sources of carbohydrates. Nature did this because its design is to bring in quick energy along with a key nutrient for the body's weight-health signalers—Switch hormones—to get and use the nutrients received as energy. The good news is that these carbohydrate sources also pack in lots of other key nutrients. There are a few non-carbohydrate-rich options to help you meet your needs. This is important because meeting your potassium needs solely from supplementation is not an optimal health strategy—not even close. Potassium supplementation needs to only be used as a support, at much lower amounts, to avoid concerns for the kidneys. Medication use will also be a consideration. The "Operator's Manual" gives you some options for increasing potassium intake from food. Look to those swaps as a way to upgrade your current choices whether you supplement or not.

Magnesium

We covered magnesium a lot in our discussion of hydration and motility. Refer back to sections on hydration and motility to get guidance on how to consider supplementation, and see the Operator's Manual for food recommendations.

Iron

Iron plays a crucial role in metabolism and the transport of oxygen throughout the body. Researchers are providing interesting insights about how it may impact the signaling for GLP-1 secretion as they also explore the potential for agonists to help optimize iron levels in some individuals. To optimize your iron levels, use better, more complete assessment. Look at total nutrition intake, what can interfere with iron absorption, what increases iron needs, and lab data that gets to the root cause of any imbalances. These questions are included in your resources that accompany the book. (See the iron evaluation and guide to better metabolic labs.) *Caution:* Optimizing iron levels can't come at the cost of other nutrient optimization. Make sure that supplementation to increase levels doesn't throw off other parts of your ecosystem, especially Switch-critical digestion.

B Vitamins, Choline, and Zinc

A power trio, B vitamins, choline, and zinc all deserve attention in their own rights but collaboratively, optimization of them is essential for Switch function, so I've grouped them together. Specifically, they are all crucial for vagus nerve function, digestive health, and optimizing methylation for cellular metabolism throughout the entire ecosystem. Getting optimal levels of B vitamins, choline, and zinc, along with magnesium and vitamin C, will support better methylation. Genetic variants can impact your body's methylation efforts; testing can provide insights to help you optimize your nutrient needs—quality and quantity.

Additionally, among zinc's numerous functions includes being a resource in the metabolic functions of the L cells, which produce and secrete GLP-1. Zinc is also active in the pancreas's beta cells, which are involved in all aspects of insulin formation and deployment.

B vitamins provide nutrient support for so many different Switch-related functions, including energy metabolism, digestive function, nerve responses, and heart health.

B12 deserves a specific callout for Switch function as it directly impacts microbes (bacteria) in the gut that promote the secretion of your Switch hormones. Digestion, specifically stomach acid, is essential for B12 to be transformed into the usable form to do its work in the small intestine. As you learned, medications—including GLP-1 agonists—can alter stomach acid and B12 utilization. Genetics, food choices, and other digestive health challenges will significantly impact your needs and how your body can use this nutrient.

Choline, on a technicality, is not a true B vitamin (it used to be called B4), so I refer to it as "cousin choline." Historically relegated to a nutrient that pregnant women should optimize, we now know better. It plays crucial roles in the body, including methylation and vagus nerve function, and its key for optimal liver health. Its importance in the human body is telegraphed by the fact that it is found in so many foods, both plant and animal sources. Yet with food processing, dietary preferences, and public health marketing messages (ditch the egg yolks), we are often insufficient in our intake. This is one you can and should optimize via food, even adding a quality source of sunflower lecithin to smoothies or baking recipes. But also assess inclusion of it in your supplementation to help optimize levels (if you are taking a multivitamin mineral, it should be included). According to my colleague Dr. Navaz Habib, the better types of choline for vagus nerve support is phosphatidyl choline or phosphatidyl serine.

Vitamin D

Vitamin D plays a direct role in your Switch function and weight-health ecosystem as it helps promote calcium and magnesium absorption via the intestines. This in turn impacts bone health as well as helps to avoid calcium traveling where it isn't meant to go. Vitamin D is also involved in thyroid and blood sugar metabolism. Your intake of vitamin D should be personalized based on labs and, in almost all cases, will require supplementation either daily or weekly.

We've been misguided to think that everyone can make enough vitamin D, known as the sunshine vitamin, with sun exposure. Skin tone, digestive health, risk factors for skin cancer (needing to avoid sun exposure), and your body's vitamin needs make this a poor recommendation for most. Though it could be part of a seasonal strategy to help maintain vitamin D levels, I recommend vitamin D supplementation for all patients where there is demonstrated insufficiency or frank deficiency. Insufficiency is a topic of much debate. My preferred range for most patients is 45–60 ng/mL for vitamin D 25 OH (this is the lab value from your assessment and the more common one tested) as a starting point. Note that for some practitioners, 60 ng/mL may be considered on the high side; for others, too low. Experts can debate whether there is an epidemic of vitamin D deficiency. There is no debate, in my opinion, for vitamin D insufficiency. Correcting for it has huge upsides for your Switch function and beyond.

Vitamin A

Many of the organs with Switch receptors rely on vitamin A compounds. Vitamin A helps activate your thyroid hormone receptors.
It operates in the gut, promoting the health of intestinal epithelial
cells and supporting microbial health—the eyes and brain. We can
and should achieve good levels of vitamin A via food choices, and
where we can, we should optimize via supplementation.

Trace Minerals

Iodine, copper, chromium, selenium, and other trace minerals
support metabolic functions throughout the body. Because they
are often a casualty of food ingredient processing, preparation,
and dietary choices, and because the damage to soil means food
is often starting with less of them, it is good to pay attention
to them.

- **Iodine** is crucial for thyroid function, along with vitamin D.
  Unfortunately these two are often not assessed and optimized
  prior to medication being prescribed, especially for low
  thyroid function. We consume less today due to soil health,
  changes in salt types—from iodized to sea salts—and with
  food processing. It should be assessed in all weight-health
  efforts and given extra emphasis when there is indication of
  suboptimal thyroid function.
- **Chromium** needs to be at optimal levels to help make cells
  available for the sugar that insulin delivers to their doors.
  Similar to iodine, assessment and optimization should occur
  for all with weight-health optimization goals, but especially

for those with suboptimal blood sugar levels. Insufficiency is largely the result of the same: soil health, food processing, and so forth.

Glucoraphanin and Myrosinase

This long-lasting antioxidant and its partner enzyme play a crucial role in your body's detoxification capabilities. Often referred to as *sulforaphane*, these compounds work together to create the sulforaphane the body needs. Guess where? In the digestive tract. Yup, it really is the hub. You can consume glucoraphanin and myrosinase from broccoli florets and sprouts, but broccoli often doesn't make the "daily" list for most of us. And there are tremendous differences in the amount depending on the broccoli seeds, soil quality, and growing environment. So, based on our needs for detoxification, it becomes a smart Switch Optimization supplement to consider.

SO THAT'S IT FOR TOPPINGS

From inflammation to digestion and detoxification, through the GLP-1 shot, GLP-1 activators and optimizers, and GLP-1–adjacent recommendations, to ingredients, we covered a lot of ground to drive personalization of your plan. Toppings can be so delicious that they feel like the reason the whole pizza works. And yet toppings by definition are location specific; they sit atop a crust, sauce, and cheese. They top them off. The whole is indeed greater than the sum or any one of the parts. Switch Optimization works just the same.

CHAPTER 13

YOUR FORCE FIELD

We defined our mission, since you did choose to accept it, as optimizing your Switch function to build your own force field and restore your weight-health ecosystem.

Has that happened?

This isn't a rhetorical or self-evaluative question.

You may be tempted to view this section as an end point. At the end of a mission, we do a post-mission report and debrief, where we evaluate the mission for success or failure, as well as its parts, to learn lessons for our next mission. That's where *The A-Team* episodes and your weight-health ecosystem's missions diverge.

We aren't ever at the end of this mission—until you are at your end. Where we are now is mid-mission.

And a great time to reassess.

Here's a piece of strategic brilliance from one of our nation's great public servants, physicist, and secretary of defense, the late Ashton Carter. He employed an approach he called "backward planning" to build missions: "When we start with the desired outcome, where we want to be, and work backward, we can plan most effectively." "So, Ashley," you may be asking, "you're telling me this now? On page three-hundred-something? Not at the start of the mission?"

Exactly.

I'm bringing this in now because what we have just been doing, this mission, is a model of backward planning. We set your goals: Optimize your Switch function, build your force field, and restore your weight-health ecosystem.

We built the mission in broad strokes. Then closer in, via assessment, and finally in the recommendations, we took it fully operational.

We could do that looking forward because all along, for years, I'd been looking back.

Backward planning gets a big assist from (re)optimization.

(Re)optimization is as vital as assessment. It's about your ROI—return on investment—which, just as in business or other areas of life, needs to be measured for our health efforts.

First, we'll see what you've implemented.

Next, we'll interpret your body's responses.

In doing so, we'll capture any improvements—we will tag those as optimizations—and identify any choices that proved to not be better because they didn't produce positive results or they were not realistically doable.

What, if anything, has changed? To what extent? What did you learn about yourself and your Switch's needs? Most specifically, where do you see directional improvement, and how and where can you see more of it?

This takes us to *your better next steps.*

YOUR (RE)OPTIMIZATION

You noted the date when you gathered your original data. Get that notebook out again so that you can run a comparison. Perhaps you already reassessed some parts of your original assessment? Great. Regardless, do the entire (re)optimization.

Once completed, here's how you compare them. In case your mindset needs to hear this, I am going to say it out loud: "You are comparing data, not changes in your personal value."

- What insights do you now have? List any changes you notice.
- What has improved, worsened, or stayed the same? *Before answering*, note that these are your observations; a data point can be a new lab, how you felt about making a choice, changes in how clothes fit, or whether something you thought you wanted to do was doable or delicious.
- Now it is time to assess your changes as follows:
  - ◇ Those you feel better about today, tag with a B.
  - ◇ Those you feel have worsened, tag with NB.
  - ◇ Those you feel stayed the same, tag with S.

These letter assignments are a starting point, not the whole story. For most of us, B's feel really good, and NB's don't. And maybe an S feels okay, or it could be like that frustrating number on the scale that didn't move. Argh. However, as data, they all provide a lot of value. In this next section, use the questions and discussion to help you evaluate the tags you've assigned to your insights to move onto your better next steps.

- Your B's: You want to keep the momentum going. How will you protect and amplify these results? Do your B's feel doable moving forward? Yes, yay, awesome sauce! If not, consider what support you will need as life happens.

- For each NB, list what you expected (or hoped) would occur and what did (or did not) occur. This should include any emotions or challenges associated with your experience ("It was too hard," "I don't have time in the mornings," etc.).
- Here are common examples of NBs and how to use the data better:
    - ◇ Life happened—to your disadvantage. Unexpected events will impact the outcome of any experiment. Don't beat yourself up about it. Is this a better time to redo that experiment or not?
    - ◇ Maybe you were a bit overambitious. Did you set a goal that was achievable? Our enthusiasm can want to be in the HOV lane, but our body really needs a consistently paced road trip, safer and dependable. Here is where we differentiate between a preset goal and the better goal of directional improvement.
    - ◇ Perhaps your body revolted. Did you attempt too many changes at one time? On occasion, doing one new thing can cause a body to respond with irritation. On most occasions, when initiating several new choices at once, the body will revolt. Not to mention, we create an impossible scenario for determining what drove specific outcomes. Hit the pause button. Let the body get over itself. And next time, choose just one.
    - ◇ Did you do the work but stress was working you over, too? If it was a really stressful time and things didn't get better, did they maybe stay the same? Then this NB might be an S when you implement the "Better, not perfect" mindset. And that *is* worth celebrating!

- What to do next with your NB's: Do you know what to experiment with next to see if it could better help your Switch function? No? Like my patients (and me), you may need to engage a practitioner to help you in this process, especially if you feel like you're not able to clearly decode your body's messages. And there's always the chance that what you learned is that what you committed to, what you experimented with by doing, or what you chose to invest in isn't a better choice for your body today. If so, it would be really good to note this down and remind yourself the next time you are spurred to invest in it or something similar.

What's an S—what stayed the same—and why? It's common, particularly after the first thirty days, to have a lot of or mostly S's. Thirty days may seem like a long time, but not for your body. The need for consistency and more time to let changes drive results are often what give us S's. Maybe we haven't been doing something as regularly as it demands—laughing once a week but not daily, drinking water with electrolytes when we work out but not on nights we drink alcohol, so we are still getting up to pee. Maybe you need to do your S's more often or . . .

- More time needs to pass: Some things can turn on your Switch really fast, and others take time. Your vagus nerve needs you to feel safe—truly safe, not faking it—for longer before it relaxes into prioritizing delivery of your Switch hormones.
- There's always the chance that what you learned is that what you committed to, what you experimented with by doing, or what you chose to invest in isn't alone going to move the

needle. What could you add to it—beyond more time or more consistency—to have it deliver optimal results?

⋄ Do you need more personalized help?

⋄ Do you need more or better data?

⋄ Do you need to try a different effort to optimize based on the results of an experiment you did?

(Re)optimization completed. You now have insights about your better next steps.

They are likely a combination of working to maintain your better (B) choices, letting go of NBs and replacing them with new experiments, and setting a new time window for when you will measure your S's. And asking for help and support where useful.

Keep going—your weight-health ecosystem is, by nature, dynamic. It's a web of living things and nonliving substances that are constantly producing stimuli and responses and interacting.

THE ONGOING MISSION BRIEFING
AND ASSESSMENT TIMING

• Daily: body and stress check-ins to identify and make your better choices

• Weekly: weight composition, hydration, sleep, and better nutrition

• Monthly: digestive

• Quarterly: weight-health reassessment, INFObesity assessment, lab data

• Additional specific reassessment: when life happens

⋄ Illness requiring treatments (over the counter, medications, therapies)

- Injury (this includes things such as dental work that impact chewing and oral health)
- Preventative testing—colonoscopy, endoscopy, scans, dental (if local anesthesia, etc.)
- Hormonal shifts
- Supplemental nutrient adjustments
- Dietary changes—giving up alcohol, becoming plant monogamous (vegan), eliminating an ingredient (gluten, dairy, etc.)
- Pregnancy, having a newborn at home
- Performances (training plans, performance of, and recovery from)
- Travel (remember Martha!)
- Life stages (starting work, changing jobs, school [yours or that of someone you care for], empty-nesting, retirement, moving, etc.)

CHAPTER 14

(RE)OPTIMIZATION

Mission success means you are not now and will never be in a rock-bottom nutrition moment like the one that led to my goat's milk cleanse. You are at a place where you understand your body's needs, can decode its signals, and can more easily navigate modern living with all it throws your way. Our work together can be ongoing—both as you come back to the book as a tool to help you navigate life's inevitable happenings and on the website as you use the existing resources and get updates.

At the start of the book, I expressed that I need you. Our world needs you. You've got a specific superpower, zone of genius, and life mission. Mine is helping you optimize it—by optimizing your weight-health ecosystem.

YOUR SWITCH OPTIMIZATION RESOURCES

Writing this book, I backed myself into a corner. I railed against INFObesity and all those contributing to it. Now I need to direct you to the internet. I want you to have all the resources you need and can use to identify, make, and evaluate the benefits of nutrition and lifestyle recommendations. My team and I will make sure those resources stay current; we will add additional ones as new hot topics and trends arise. You have our word. We will not contribute to INFObesity—out in the world or in your life. To that end, you are also getting access to humans, live and recorded, to help you vet your INFOload and support you as you conduct experiments.

Too much knowledge is not power. But insights about your body and its needs are key.

PART VI

OPERATOR'S MANUAL

HOW TO FISH BETTER

Let's go back to fly-fishing for another story that should help you navigate these recommendations. Remember the calm, cool, collected vagus nerve whisperer who happens to be my brother? He taught me to cast. Where others had failed, he succeeded. One of the flops was a guy I was sort of dating; the other was a friendly Canadian with a ten-and-two approach. The friend I was in the mountains with got it. I didn't. As we know, I'm competitive, so I persisted, which didn't help. I was, as the guide noted, nine-to-fiving it, not ten-and-two-ing—not Brad Pitt in *A River Runs Through It*; more Edward Scissorhands in *Edward Scissorhands*. The poor guy even drew a clock in the sand in case I was only familiar with digital watches.

Hearing about my failed attempts, my brother laughed and said something I've been borrowing ever since: "I teach my guiding students that the same reference won't work with everyone. They can find one that clicks for each person who comes out." We arrived at an arm rhythm that resembled my butterfly stroke from my swimming days—it probably makes no sense to you. That's cool. Doesn't matter. It's what worked for me. That day I caught a really, really big fish. My third time on a fly-fishing trip, I got the hang of casting. And I've gotten better at it since. Better, not perfect.

What bigger fish can we want to land than better weight health? What I learned from my fly-fishing-guide brother and my patients is that we all learn differently. While I might be a great guide for you in this book, you may benefit from hearing and seeing things differently. That's why I've added access to additional resources to support you moving forward.

SWITCH ASSESSMENT PROCESSING RESULTS WORKSHEET

PART 1

- Use your results from the quiz in Chapter 10.
- With the guidance shared, check the box(es) that applies to your results.
- After filling out the grid below for all the sections, you will have what you need to process your results.

Section	Optimal	Suboptimal	Suppressed	Delayed	Dysfunctional
A feelings					
B breathing			XXXXX	XXXXX	XXXXXXX
C labs					
D digestion hydration					
E weight composition			XXXXX	XXXXX	XXXXXX

Section A

Suboptimal or optimal? How many of each letter do you have?

____ A.

____ B.

____ C.

____ D.

_____ Total (A+B+C)

Is the total for your As, Bs, and Cs more than your total Ds?

- Yes? Record suboptimal function.
- No? Record optimal function.

For those with the suboptimal function result, let's get more details. Look at the letters you picked and select which of the below applies:

- If you have 2+ A answers, your current Switch-function status starts off as suboptimal-dysfunctional.
- If you have 2+ B answers *or* if you answered one of each (A, B, C), your current Switch-function status starts off as suboptimal-delayed.
- If you have 2+ C answers, your current Switch-function status starts off as suboptimal-suppressed.

Section B

Insert your data: Add up your yeses and nos.

_____ YES

_____ NO

All yeses? Then no part of this section suggests you have suboptimal Switch status. Record your answer to this section as *optimal.*

Any nos? If any of your results in this section are a no, your current Switch status is *suboptimal.*

Section C

Insert your data: Add up your yeses and nos.

_____ YES

_____ NO

If you answered yes to *all* of the data points in this section, or at minimum if you answered yes to *all* the ones you have (at minimum the five core), there's good objective data that your Switch status is optimal based on bloodwork lab data. That's all you record for this section.

If you answered any nos in this section (at minimum the five core), your Switch is expressing some degree of suboptimal Switch function.

If your data includes only one no (the rest are yeses), then your status is more likely suboptimal-delayed. *Note:* We do not use labs alone for a reason, so how close or far your one lab is to optimal will not be a factor here. We will address that in the personalization of your plan.

If your data includes more than one no, then your Switch function is suboptimal-dysfunctional.

Section D

Insert your data here from hydration and digestion.

_____ YES

_____ NO

If you answered all nos, then record *optimal* for this section.

If you answered yes to any of the questions, you have a suboptimal Switch function today.

- If you answered yes to 2+ questions, your suboptimal function is dysfunctional.
- If you had only one yes, your answers here indicate suboptimal-delayed status.
- If you answered yes to the hydration experiment, your Switch is suboptimal.

Section E

Insert your data here from weight composition assessment.

_____ YES

_____ NO

If you answered yes to *all* of these questions, record *optimal*.

If you answered no to *any* of these questions, record *suboptimal*.

PART 2

- Review your results from above (or copy them here for easier reference).
- Use the guidance below to determine your status and, if suboptimal, your status type.

Section	Optimal	Suboptimal	Suppressed	Delayed	Dysfunctional
A **feelings**					
B **breathing**			XXXXX	XXXXX	XXXXXXX
C **labs**					
D **digestion** **hydration**					
E **weight** **composition**			XXXXX	XXXXX	XXXXXX

Interpreting the Results

If your answer for each section was *optimal*, that is your current Switch-function status.

If you have *suboptimal* as an answer for *any* of the sections, your current Switch-function status is suboptimal. Proceed to the next section to determine your phase 2 suboptimal status.

What Type of Suboptimal?

Count up the number of answers you have for each type of suboptimal function: how many suppressed, how many delayed, how many dysfunctional.

The one you have the most of is your current type of suboptimal Switch function. Review this for clarity and how to handle a tie.

- If they are all dysfunctional, your status is dysfunctional.
- If they were all delayed, your status is delayed.
- If they were some delayed and some dysfunctional, your status is tied between the two. You can note this as dysfunctional (but closer to delayed).
- If they were only suppressed—meaning all others were optimal, but section A was suboptimal-suppressed—then your current function is suboptimal-suppressed.
- If some were delayed or dysfunctional and your section A answer is suppressed, your status is suboptimal-delayed (but closer to suppressed).

YOUR CURRENT SWITCH-FUNCTION STATUS

Phase 1 result:	Date:
Phase 2 result:	Date:

SWITCH OPTIMIZATION NUTRITION PLAN RECOMMENDATIONS

Use this plan as you evaluate insights about your current choices from your assessment. Review "Getting Started" to recap the pillars and guidelines, then move into the plan's recommendations and use the resource lists provided to help you identify and make choices that match the goal of your experiment(s).

GETTING STARTED

Thinking about your current choices and your assessment results, what stands out right away as "Oh no" or "Uh-oh"?! Exhale and remember "Better, not perfect." Our goal is to experiment with choices that help you optimize your body's responses, not to eliminate your favorites and start eating things that aren't delicious to you.

Note: Anything in the guidelines that is different from what your practitioner recommended for you should be discussed with them before you make new ones.

NUTRITION PLAN RECOMMENDATIONS

Let's recap the pillars:

- Quantity and nutrient balance: Trial balancing and consuming amounts at each pit stop (meals and snacks) that are in the ranges shared: carbs (15–30 grams), protein (15–30 grams), fats (5–15 grams), fiber (more than 7 grams), and unlimited non-starchy vegetables. These amounts apply to your total for your food, supplements, and beverages at that pit stop.

- Timing: Trial limiting your calories (including those from supplements) to a ten-to-twelve-hour window (such as 8 a.m.–6 p.m. or 7 a.m.–7 p.m.), aiming to start your calories by 10 a.m. and finish by 8 p.m. During your caloric window, pit-stop for nutrition about every three hours.

- Quality: Consume foods, beverages, and supplements that provide food in as close as possible to its natural form. If it is processed, is the processing something you could do in your kitchen (with the tools, skills, and desire)? Or must it be made in a chemistry lab? For those products and ingredients, reduce your intake, trying to limit or avoid artificial colors, sweeteners, flavors, and preservatives—for example: AminoSweet/aspartame, sucralose, phenylalanine, caramel, blue nos. 1 and 2, green no. 3, red nos. 3 and 40, yellow no. 6, tartrazine, sodium nitrate/nitrite, BHA, BHT, potassium bromate, brominated vegetable oil (BVO), BPA, and carrageenan.

- Fiber: The amount and types of fiber you take in daily will play a key role in GLP-1 production. Aim to take in more than 7 grams of fiber at each pit stop. Experiment with different sources of fiber to get a diversity of types (more guidance in the additional resources online). However, use your assessment and recommendations to personalize your fiber types, amounts, and timing.

- Colors: Aim for a daily rainbow of colors from a variety of vegetables, fruit, herbs, and spices.
- Switch optimizers: Aim for at least one serving daily of Switch Optimization promoting foods and beverages (choose organic wherever possible).
  - ◇ Cruciferous vegetables (broccoli, broccoli sprouts, cauliflower, brussels sprouts, cabbage, arugula)
  - ◇ Citrus, citrus peel (lemons, limes)
  - ◇ Resistant starches: green bananas, plantains, oats, lentils, chickpeas, red kidney beans, potatoes, barley, brown rice
  - ◇ Leafy greens (kale, arugula, dandelions, radicchio, chicory)
  - ◇ Cacao/cocoa, coffee, green teas
  - ◇ Blueberries, pomegranates, beets, plums, cherries, apples, strawberries, black currants, olives, turmeric
  - ◇ Extra-virgin olive oil, cold-pressed
- Amino acids: As you make your protein choices, see if your choices get you to 3–5 grams of glutamine and leucine, and more than 500 milligrams of taurine—at several pit stops. Adjust your supplementation based on what you don't take in from food regularly.
- Omega-3 fatty acids: Aim for about 3,000 milligrams (3 grams) daily from foods such as nuts and seeds (walnuts, hemp, flax, chia), fish (salmon, sardines, anchovies, halibut, mackerel), beef, low- and full-fat dairy, soybeans, and vegetables (brussels sprouts, broccoli, and cauliflower); and supplements as needed.
- Potassium: Aim for more than 3,000 milligrams (3 grams) daily from foods including avocado, banana, potato/sweet

potato, cauliflower, chickpeas, spinach, broccoli, beet greens, lentils, coconut water; and supplements as needed.

- Magnesium: Aim for about 600 daily from foods including nuts (cashews, almonds, peanuts), seeds (hemp, pumpkin, sesame, sunflower), cacao nibs/powder, dark chocolate (higher than 70 percent), greens (spinach, Swiss chard, beet greens, turnip greens), whole grains (oatmeal, brown rice, quinoa, buckwheat, millet), beans/legumes (kidney, pinto, black, navy, lima, tempeh, edamame); and supplements as needed. (Depending on your current intake, you may start with a goal of 400 milligrams and then increase as tolerated.)

- Added salt or sodium: Aim for a maximum of 1.5 teaspoons (about 3,450 milligrams); sources include iodized sea salts, packaged foods and beverages, restaurants food/takeout, and supplements. If consuming more, evaluate the impact on your ecosystem and work to ensure balance with other electrolytes.

- Sweeteners: Aim for less than 2 teaspoons daily of added sugar (total from foods and supplements), where 1 teaspoon is about 4 grams of sugar on a label. And no more than one serving of "natural" nonnutritive sweeteners such as stevia, monk fruit, allulose, erythritol, or xylitol. Use the Sweet Taste Bud Test (and Reset) to optimize.

- Caffeine: Evaluate impact on digestion, sleep, energy; evaluate the quality of your sources, reducing those that do not meet quality standards.

- Water: Aim for more than 40 ounces of what is listed as quality water options; spread throughout your day; use the Hose or Sponge? hydration experiment to optimize.

- Alcohol: Consider avoiding for thirty days or aim for less than 2 drinks per week.

NUTRIENT LISTS

RECOMMENDED DAILY TOTAL NUTRIENT INTAKE LEVELS (FOODS AND SUPPLEMENTS)

These are the starting points from which to use data and experimentation to optimize for human adults. Everyone should personalize. Personalization as we discuss in the book is far beyond your gender and age, today.

NUTRIENT	TOTAL AMOUNT PER DAY
Vitamin A	900 mcg
B1(thiamine)	1.2 mg
B2	1.2 mg
B3	15 mg NE (niacin equivalents)
B5	5 mg
B6	1.3 mg
B7 (biotin)	40 mcg
B9 (folate, not folic acid*)	400 mcg
B12	2.4 mcg
Vitamin C	150 mg
Vitamin D	25 mg (1,000 IU)
Vitamin K	100 mcg
Magnesium	600 mg daily
Calcium	500–1,000 mg daily
Zinc	12 mg
Iron	8 mg (men); 18 mg (women with menses)
Copper	900 mcg
Chromium	40 mcg
Potassium	3,000 mg daily
Sodium	1 teaspoon (2,300 mg)
Selenium	100 mcg
Iodine	150 mcg

Protein (About 15 Grams)

* Good source of leucine (greater than equal to 1 gram)

** Good source of taurine (greater than equal to 200 milligrams)

*** Good source of glutamine (greater than equal to 1 gram)

PLANT-BASED PROTEINS

- Black beans (cooked)—about ¾ cup
- Brown rice (cooked)—3 cups
- Chickpeas (cooked)—about ¾ cup
- Kidney beans (cooked)—1 cup
- Lentils (cooked)—about ¾ cup *
- Navy beans (cooked)—1 cup
- Oatmeal (cooked)—2½ cups
- Pinto beans (cooked)—about ¾ cup
- Quinoa (cooked)—1⅔ cups
- Soybeans (cooked)—about ½ cup ***
- Soy milk—2 cups
- Tempeh—about ⅓ block (4 ounces) *
- Tofu (firm)—⅓ cup *
- Whole wheat bread—6 slices

DAIRY AND EGGS

- Cheddar cheese—about 2 ounces
- Cottage cheese—⅔ cup (low fat) or ½ cup (full fat) *
- Eggs, large—about 3 eggs *
- Milk, low fat/2 percent or whole/full fat—about 16 ounces
- Mozzarella (soft cheese)—2½ ounces
- Parmesan (hard cheese)—1½ ounces
- Yogurt, Greek (plain)—1 cup (low fat) or ¾ cup (full fat) *

SEAFOOD

- Anchovies—2 ounces
- Halibut—about 3 ounces ***
- Lobster, cooked—about 3 ounces
- Mahi mahi—about 3 ounces
- Salmon, cooked—about 3 ounces ***
- Sardines—about 2 ounces **
- Scallops, baked—5–7 large (4 ounces) **
- Shrimp—about 3 ounces ***
- Snapper—about 2 ounces *
- Tuna, canned in water—about 2 ounces ***

ANIMAL PROTEINS***

- Beef, ground, 90 percent—about 2 ounces
- Beef liver (cooked)—about 2 ounces
- Bison (cooked)—about 3 ounces
- Brisket—about 3 ounces
- Chicken breast (cooked)—about 3 ounces
- Chicken (dark meat, cooked)—about 2 ounces **
- Duck (cooked)—about 3 ounces
- Ham (lean)—about 3 ounces
- Ostrich—about 2 ounces *
- Pork roast—about 2 ounces *
- Sirloin steak—about 2 ounces
- Turkey (cooked)—about 2 ounces **
- Turkey (dark meat, cooked)—about 2 ounces **
- Venison—about 2 ounces *

NUTS AND SEEDS

- Almonds—½ cup
- Cashews—¾ cup

- Chia seeds—about 6 tablespoons
- Flax seeds—½ cup
- Hemp seeds—about 5 tablespoons *
- Peanut or almond butter—4 tablespoons (¼ cup) *
- Peanuts—½ cup *
- Pecans—1½ cups
- Pumpkin seeds (pepitas)—⅓ cup *
- Sunflower seeds—⅔ cup
- Walnuts—1 cup

Vitamin C (About 100 Milligrams)

- Bell pepper, red—½ cup, raw
- Broccoli—1 cup, cooked
- Brussels sprouts—¾ cup, cooked
- Cantaloupe—¼ cup
- Cauliflower—1 cup, cooked
- Kale—2 cups, cooked
- Kiwi—1 large
- Mango—¾ cup
- Orange—1 medium
- Papaya—¾ cup
- Pineapple—¾ cup
- Strawberries—1 cup

Magnesium (About 100 Milligrams)

NUTS AND SEEDS

- Almonds—2 tablespoons
- Cacao nibs/powder—4 tablespoons
- Cashews—3 tablespoons
- Chia seeds—1 tablespoon

- Chocolate, dark (higher than 65 percent)—2 ounces
- Hemp seeds—1 tablespoon
- Peanuts—⅓ cup
- Pumpkin seeds—⅓ cup
- Sesame seeds—3 tablespoons
- Sunflower seeds—¾ cup

GREENS

- Avocado—½ avocado
- Beet greens—1 cup, cooked; 4 cups raw
- Spinach—¾ cup, cooked; 4 cups raw
- Swiss chard—¾ cup, cooked; 4 cups raw
- Turnip greens—1 cup, cooked; 4 cups raw

WHOLE GRAINS

- Brown Rice—1 cup, cooked
- Buckwheat—1 cup, cooked
- Millet—1 cup, cooked
- Oatmeal—1½ cups, cooked
- Quinoa—1 cup, cooked

BEANS AND LEGUMES

- Black beans—½ cup, cooked
- Edamame—1 cup
- Kidney beans—½ cup, cooked
- Lima beans—1 cup, cooked
- Navy beans—1 cup, cooked
- Pinto beans—½ cup, cooked
- Tempeh—1 cup

Potassium (About 300 Milligrams)

VEGETABLES/GREENS

- Acorn squash—¾ cup, cubed and cooked
- Asparagus—¾ cup, cooked
- Beet greens—¾ cup, cooked
- Beets—¾ cup, cooked
- Bok choy—2 cups, cooked
- Broccoli—¾ cup, cooked
- Brussels sprouts—¾ cup, cooked
- Cabbage—¾ cup, cooked
- Carrots—¾ cup, cooked
- Mushrooms—1 cup, cooked
- Potato, sweet or white—½ cup, cooked
- Spinach—¾ cup, cooked
- Summer squash—¾ cup, cubed
- Swiss chard—2 cups, cooked
- Tomatoes—¾ cup, cooked
- Turnip greens—2 cups, cooked

BEANS AND LEGUMES

- Lentils—½ cup, boiled
- Lima beans—½ cup, boiled
- Pinto beans—¾ cup, boiled
- Soybeans—½ cup, boiled
- White beans—¼ cup, boiled

ANIMAL PROTEINS

- Cod—2½ ounces
- Salmon—2½ ounces
- Tuna—3 ounces

FRUITS AND SEEDS

- Apricots, dried—¾ cup, sliced
- Avocado—½ cup, sliced or mashed
- Banana—1 medium
- Cantaloupe—1½ cups
- Coconut meat—1½ cups, shredded
- Kiwi—1 medium
- Peaches, dried—1 whole
- Prunes—3–4 pieces
- Pumpkin seeds—¼ cup, unsalted, roasted
- Raisins—2 tablespoons
- Watermelon—2 cups

BEVERAGES

- Carrot juice—3 ounces, fresh squeezed
- Coconut water—4½–6 ounces
- Orange juice—½ cup, fresh squeezed
- Tomato juice—4½ ounces, fresh squeezed
- Watermelon water—6 ounces

SWITCH EXPERIMENTS

THE SWEET TASTE BUD TEST

You will need a few bites of an apple (a quarter apple or two slices). With the exception of a green Granny Smith or candied or baked apple, which are not allowed, the type does not matter that much. Choosing one that isn't advertised for being super sweet is preferred. It's more important to use the same apple type—and about the same degree of ripeness—for your test and retest so that you get accurate results. If you cannot consume apples, you can use a pear instead.

Consider previous bites and sips that are sweet to you and mentally develop a scale from 1 to 10, where 1 is not sweet at all and 10 is very sweet, the most sweet that you can recall.

With a neutral palate (not recently following tooth brushing, using mouthwash, or consuming coffee, alcohol, food, gum, or mints—if you have, eat a few slices of cucumber or sip on some water to cleanse your palate, take two to three bites of your apple, chewing thoroughly.

As you chew and experience the flavor, score your bites on your scale from 1 to 10.

On average, what do your bites score?

Results

- If your apple bites are greater than a 7, your sweet taste buds appear to be well-aligned with the natural degree of sweetness.
- If your apple bites are a 6 or less, you will want to conduct a reset and retest your apple bites every seven to ten days until you achieve a score of 7 or greater.

THE SWEET TASTE BUD RESET

Remember that this is for a set period of time, not a life sentence.

Remove from your diet all added sugar and sweeteners—sugars, syrups, honey, and so forth—as well as ready-made products containing them. Also avoid all nonnutritive sweeteners . . . yes, even the natural ones.

If your current supplements—including fiber and protein powders—contain sweeteners or added sugar, work with your practitioner or make replacement choices for the duration of the reset. You may be onto something with your sweet taste buds if you discover these as a source!

Yes, you can have foods that have naturally occurring sugars such as fruits and vegetables. Incorporating them into your overall nutrition plan may really help you make it through the reset! Baking or preparing them with a pinch of salt or spices such as cinnamon may help bring out their sweetness to help satisfy you. You can also consume herbal teas that have a natural sweetness.

During your reset, document and experiment with strategies to address when you experience sweet cravings. Are you tired? Hungry? Thirsty? Are your pit stops nutrient balanced and better timed, and do they include quality ingredients in better quantities?

Did you consume something high in salt? Something not delicious to you in the moment? What about post caffeine or alcohol? How is your digestion—gassy, bloated, constipated, or loose stools?

These insights will help you optimize your sweet taste buds and your weight-health ecosystem using the discussion points and additional experiments in the book. You may need some time for those experiments—such as adding a probiotic—to impact your system before you see improvements in your sweet taste buds.

Redo the Sweet Taste Bud Test after seven to ten days, and continue if you are not yet at a 7. When you achieve a 7, you can experiment with bringing back choices such as added sugar and sweeteners. Informed by your insights, experiments, and the reset experience, the amounts that you add back should help you maintain optimal sweet taste bud function.

Retest once a quarter or when you notice you are challenged by sweet cravings.

CONDUCTING A VITAMIN C FLUSH

Considerations: One way to determine how much vitamin C from supplementation will be beneficial especially to improve bowel motility and elimination is known as a vitamin C "flush." When motility challenges persist despite optimizing magnesium (and breathing and movement), I use this as a complementary tool to establish the amount of vitamin C at which the body will eliminate. When at its maximum, it will "flush" the bowels— and yes, with urgency—eliminating via watery or loose stool. From that maximum, we can determine a better level to start using, and then adjust accordingly. This requires time at home and proximity to a bathroom, ideally with as little elevated stress as possible. It typically takes about four hours. The remainder of the day and the following day, the bowels may still be loose and you may feel lower in energy. Avoid having plans—especially those with travel, alcohol, intense activity, or limited access to a bathroom. Also note that this flush can throw off CGM (continuous glucose monitor) data.

Caution: Better to conduct a vitamin C flush under the guidance of your practitioner. The night before your flush, they may adjust your food and beverage choices, and any medications and supplements during the flush. Do not do it if you are pregnant, undergoing any type of medical treatment, or have any digestive disease. Your practitioner should guide you in the selection of your vitamin C supplement. I recommend a buffered l-ascorbate as a powder.

Instructions

Step 1: On waking with an empty stomach, consume 2–4 ounces of water while taking your first dose of vitamin C (as advised or 500 milligrams). Record the timing.

Step 2: Every thirty minutes, repeat step 1. Stop when you experience your first watery stool.

Step 3: The amount of vitamin C you took when you had loose stool is your maximum; the number you want is about 75 percent of your max. So if your maximum is 4,000 milligrams, you will multiply that by .75 to get 3,000 milligrams. This will be your starting daily dose. *Make sure to consume it in divided doses the following day and forward.* Moving forward, if you notice your stool is staying loose, reduce your vitamin C further.

ASSESSING YOUR POOP QUALITY— BRISTOL STOOL FORM SCALE

BRISTOL STOOL FORM SCALE

Type 1		Separate hard lumps, like nuts
Type 2		Sausage-shape, but lumpy
Type 3		Like a sausage but cracks on the surface
Type 4		Like a sausage or snake, smooth and soft
Type 5		Soft blob with clear-cut edges
Type 6		Fluffy pieces with ragged edges, a mushy stool
Type 7		Watery, no solid pieces

THE INFOBESITY PROTOCOL: HOW TO IDENTIFY AND ADDRESS INFOBESITY

There is a *better* way to consume health information as well as determine what should stay as entertainment versus what can be actionable.

The INFObesity Assessment

The INFObesity assessment is a tool to review and adjust your INFOload and sources. Use it as often as needed, but at least once every season (ninety days)—just like you change your clothes or food choices.

Step 1: Identify your INFOload.

Identify the source(s) of information to which you are exposed. Jot them down on a piece of paper or as a note on your phone. Don't hold back. Yes, the chat with your friend, a stranger, that dinner party guest, your fitness instructor; books, book reviews, podcasts, social media, online ads, emails, and spam.

Step 2: INFOvetting

Consider how much these sources know about you, what your body needs *today*, your specific health issues and concerns, your personal values, and why you make the nutrition and lifestyle choices you currently make.

Looking at your sources, put a D next to any daily information sources, an O next to information sources that you turn to often, and an I next to sources that influence your investments (what you spend money on).

Using the graphic below, map your sources.

- In the bull's-eye, insert your sources that *truly* know you. They know who you are and what is most important to you. They know your why and your purpose for your health and wellness goals.
- Next, think about the sources of information that know or share aspects of you but they don't know all about you and your body, today. For example, they could know you are a vegetarian woman who is going to have a child. Or they asked you questions about your food choices and health issues or concerns. Place these sources of information in the second circle of the bull's-eye.
- The last circle is dedicated to the sources of information that don't know you at all. They are sharing headlines or full research studies for which you were not a subject. These sources are sharing their own or their patients' information and recommendations for them. Maybe the person inspires you or you want to achieve the results they have shared.

Step 3: Competing your INFOassessment

As you look at your circles, evaluate the quantity of the sources in each one. Here are some suggested actions to take based on those findings.

Do you have more than two sources with a D? *If yes, reduce until you only have two. Even if better quality, too much is just too much. You can assign different D's to different days if you want more variety.*

Do you have more than five sources with an O? *If yes, reduce until you only have five. Same as the note above—go for quality and quantity more often.*

Quality sources that serve your wellness; where you should invest your time, money and emotions.

Information sources that may provide beneficial insights for your health but requires additional validation.

Information sources that don't help & may challenge your health; consider elimination.

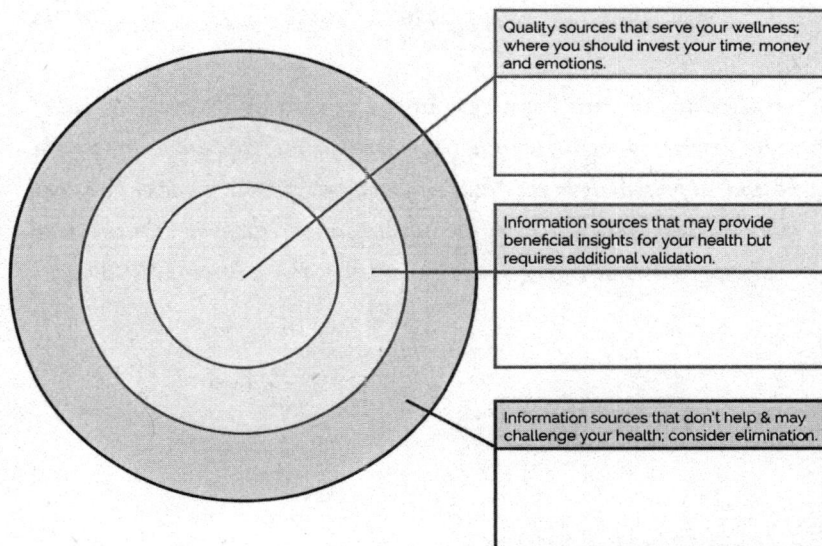

OUTSIDE CIRCLE	INNER CIRCLE	BULL'S-EYE
Information sources that don't know you personally so recommendations will not be personalized to you and your body today; consider elimination.	Information sources that may provide beneficial insights for your health but are not able to truly personalize for you and your body today; require additional validation.	Quality sources that should be able to support you with truly personalized recommendations to identify what's better and not better for your body; likely the better investment of your time, money, and emotions.

Do you have more than five sources in your bull's-eye? *If yes, this can be great that you have so much truly personalized support. However, more in your bull's-eye increases the risk of a "too many cooks in the kitchen" issue. Get clear on the value of each one, right now. Do you share your current health insights, recommendations, and efforts*

346 YOUR BEST SHOT

among them all? (Is the team playing from the same playbook, with the same intel?)

Are any of your I sources outside of your bull's-eye? *Get rid of these. Investing resources including money, time, and emotions should be limited primarily to your bull's-eye sources. If there is one or two that you can't part with, and they are in your inner circle, you may choose to evaluate your investment but focus on the goal—a better INFOload.*

ACKNOWLEDGMENTS

I am a part of an incredible ecosystem that helps me fulfill my life's purpose. Acknowledging all the participants within this ecosystem individually isn't possible here, so this is intentionally impersonal. Starting with "that guy" in the bar who asked me about antibiotics and, yes, even "that woman" who diagnosed my worm, along with all my practitioners before and after—thank you for being my teachers, helping me on my quest to understand my body and identify the choices that make it run better. Borrowing from my niece and nephew who call me the "backwards DR," as an RD I am beyond grateful for the time, enthusiasm, and support of so many amazing doctors, only a few of whom are mentioned here, beginning with the incomparable Dr. Tieraona Low Dog—you role model fulfilling your life's purpose with grace, humor, and fun. I am so grateful for every part of your journey that you bring me on! Dr. Weil, Dr. Khalsa, Dr. Merrell, and Dr. Bland, thank you for dedicating your lives to provide frameworks to change how health care is viewed and delivered, and for your support. My friends and my colleagues—many of you are both—every moment we get together is a gift. Thank you for supporting, loving, laughing with, and enabling me to course correct as needed. Myles, I know we could have hit the Bourbon trail instead of Everesting; thank you forever for saying yes to that pivot and all that ensued. Steph and Tess, I love you both so much and in this case for pushing me to take my "best" shot, where better wouldn't do. Thank you to every coffee shop owner who put up with me needing espresso and Wi-Fi real early. To my countless coaches, thank you for helping me build better muscles figuratively and literally, and helping me learn when to flex and when to relax them. My team at BNP—Anne, Ashley, Nicole, Karen, Lara, Kathy and so many more—you are the arrows to my bull's-eye and have such a special place in my heart. Thank you Sharon and Jess, for saying yes, for being the best teammates, for upleveling this work in every way, for doing it with the best senses of humor, and for protecting it and me every step of the way. Amy Stanton you are a unique force, who I am so lucky, on repeat, to have in my corner; thank you for creating a team that supports, pushes (and pushes back), and drives me to co-create collaborations that can help change the world for the better.

To my family, those here and those who've passed, I love you. Rerun and Roger, you've expanded my heart and made it possible for me to meet so many incredible people and dogs, I'm grateful to you both every day.

Lastly, I want to acknowledge you and your investment as you take action, using this system to optimize yours. Thank you. I can't wait to see how our ecosystem becomes even better as you optimize your weight health on your path to fulfilling your life's purpose.

REFERENCES

CURRENT STATE OF US HEALTH

Almario, Christopher V., Megana L. Ballal, Willaim D. Chey, Carl Nordstrom, Dinesh Khanna, and Brennan M. R. Spiegel. "Burden of Gastrointestinal Symptoms in the United States: Results of a Nationally Representative Survey of Over 71,000 Americans." *American Journal of Gastroenterology* 113, no. 11 (November 2018): 1701–10. https://doi.org/10.1038/s41395-018 -0256-8.

Araújo, Joana, Jianwen Cai, and June Stevens. "Prevalence of Optimal Metabolic Health in American Adults: National Health and Nutrition Examination Survey 2009–2016." *Metabolic Syndrome and Related Disorders* 17, no. 1 (February 2019): 46–52. https://doi.org/10.1089/met.2018.0105.

Khoshbin, Katayoun, and Michael Camilleri. "Effects of Dietary Components on Intestinal Permeability in Health and Disease." *American Journal of Physiology: Gastrointestinal and Liver Physiology* 319, no. 5 (November 2020): G589–G608. https://doi.org/10.1152/ajpgi.00245.2020.

SWITCH OPTIMIZATION NUTRIENTS AND NUTRITION

Magnesium

Rosanoff, Andrea. "Perspective: US Adult Magnesium Requirements Need Updating: Impacts of Rising Body Weights and Data-Derived Variance." *Advances in Nutrition* 12, no. 2 (March 2021): 298–304. https://doi .org/10.1093/advances/nmaa140.

Fatty Acid (C15:0)

Venn-Watson, Stephanie, Richard Lumpkin, and Edward A. Dennis. "Efficacy of Dietary Odd-Chain Saturated Fatty Acid Pentadecanoic Acid Parallels Broad Associated Health Benefits in Humans: Could It Be Essential? *Scientific Reports* 10 (2020). https://doi.org/10.1038/s41598-020-64960-y.

PRO-RESOLVING MEDIATORS

Basil, Maria C., and Bruce D. Levy. "Specialized Pro-Resolving Mediators: Endogenous Regulators of Infection and Inflammation." *Nature Reviews Immunology* 16 (2016): 51–67. https://doi.org/10.1038/nri.2015.4.

VITAMIN C

Levine Mark, Yaohui Wang, Sebastian J. Padayatty, and Jason Morrow. "A New Recommended Dietary Allowance of Vitamin C for Healthy Young Women." *PNAS* 98, no. 17 (August 2001): 9842–46. https://doi.org/10.1073/pnas.171318198.

Taurine

Wójcik, Oktawia P., Karen L. Koenig, Anne Zeleniuch-Jacquotte, Max Costa, and Yu Chen. "The Potential Protective Effects of Taurine on Coronary Heart Disease." *Atherosclerosis* 208, no. 1 (January 2010): 19–25. https://doi.org/10.1016/j.atherosclerosis.2009.06.002.

Leucine

Chen, Qixuan, and Raylene A. Reimer. "Dairy Protein and Leucine Alter GLP-1 Release and mRNA of Genes Involved in Intestinal Lipid Metabolism In Vitro." *Nutrition* 25, no. 3 (March 2009): 340–49. https://doi.org/10.1016/j.nut.2008.08.012.

GLUTAMINE PROMOTES GLP-1 AND GIP

Greenfield, Jerry R., I. Sadaf Farooqi, Julia M. Keogh et al. "Oral Glutamine Increases Circulating Glucagon-Like Peptide 1, Glucagon, and Insulin Concentrations in Lean, Obese, and Type 2 Diabetic Subjects." *American Jour-*

nal of Clinical Nutrition 89, no. 1 (2009): 106–13. https://doi.org/10.3945 /ajcn.2008.26362.

BITTER TASTE RECEPTORS

Jalševac, Florijan, Ximena Terra, Esther Rodríguez-Gallego et al. "The Hidden One: What We Know About Bitter Taste Receptor 39." *Frontiers in Endocrinology* 13 (March 2022). https://doi.org/10.3389/fendo.2022.854718.

Yu, Yunli, Gang Hao, Quanying Zhang et al. "Berberine Induces GLP-1 Secretion Through Activation of Bitter Taste Receptor Pathways." *Biochemical Pharmacology* 97, no. 2 (2015): 173–77. https://doi.org/10.1016/j .bcp.2015.07.012.

SWEETENERS

Gauthier, Ellie, Fermin I. Milagro, and Santiago Navas-Carretero. "Effect of Low- and Non-Calorie Sweeteners on the Gut Microbiota: A Review of Clinical Trials and Cross-Sectional Studies." *Nutrition* 117 (2024). https:// doi.org/10.1016/j.nut.2023.112237.

Suez, Jotham, Yotam Cohen, Rafael Valdés-Mas et al. "Personalized Microbiome-Driven Effects of Non-Nutritive Sweeteners on Human Glucose Tolerance." *Cell* 185, no. 18 (August 2022): 3307–28. https://doi .org/10.1016/j.cell.2022.07.016.

GLP-1 FOODS

Blueberries

Curtis, Peter J., Lindsey Berends, Vera van der Velpen et al. "Blueberry Anthocyanin Intake Attenuates the Postprandial Cardiometabolic Effect of an Energy-Dense Food Challenge: Results from a Double Blind, Randomized Controlled Trial in Metabolic Syndrome Participants." *Clinical Nutrition* 41, no. 1 (2022): 165–76. https://doi.org/10.1016/j.clnu.2021.11.030.

Kato, Masaki, Tsubasa Tani, Norihiko Terahara, and Takanori Tsuda. "The Anthocyanin Delphinidin 3-Rutinoside Stimulates Glucagon-Like Peptide-1 Secretion in Murine GLUTag Cell Line via the Ca2+/

Calmodulin-Dependent Kinase II Pathway." *PLoS ONE* 10, no. 5 (2015): e0126157. https://doi.org/10.1371/journal.pone.0126157.

Li, Rui, Shumeng Du, Zhan Ye, Wei Yang, and Yuanfa Liu. "Blueberry Anthocyanin Extracts (BAEs) Protect Retinal and Retinal Pigment Epithelium Function from High-Glucose-Induced Apoptosis by Activating GLP-1R/ Akt Signaling." *Journal of Agricultural and Food Chemistry* 73, no. 10 (2025). https://doi.org/10.1021/acs.jafc.4c08978.

Green Tea

Ito, Aoi, Yuji Matsui, Masao Takeshita, Mitsuhiro Katashima, Chiho Goto, and Kiyonori Kuriki. "Gut Microbiota-Mediated Associations of Green Tea and Catechin Intakes with Glucose Metabolism in Individuals Without Type 2 Diabetes Mellitus: A Four-Season Observational Study with Mediation Analysis." *Archives of Microbiology* 205, no. 5 (April 2023). https://doi.org/10.1007/s00203-023-03522-y.

Jia, Ming-Jie, Xing-Ning Liu, Ying-Chao Liang, De-Liang Liu, and Hui-Lin Li. "The Effect of Green Tea on Patients with Type 2 Diabetes Mellitus: A Meta-Analysis." *Medicine* 103, no. 47 (2024): e39702. https://doi.org/10.1097/MD.0000000000039702.

Liu, Chia-Yu, Chien-Jung Huang, Lin-Huang Huang, I-Ju Chen, Jung-Peng Chiu, and Chung-Hua Hsu. "Effects of Green Tea Extract on Insulin Resistance and Glucagon-Like Peptide 1 in Patients with Type 2 Diabetes and Lipid Abnormalities: A Randomized, Double-Blinded, and Placebo-Controlled Trial." *PLoS ONE* 9, no. 3 (2014): e91163. https://doi.org/10.1371/journal.pone.0091163.

Xu, Renfan, Yang Bai, Ke Yang, and Guangzhi Chen. "Effects of Green Tea Consumption on Glycemic Control: A Systematic Review and Meta-Analysis of Randomized Controlled Trials." *Nutrition & Metabolism* 17, no. 1 (2020). https://doi.org/10.1186/s12986-020-00469-5.

Extra-Virgin Olive Oil

Bartimoccia, Simona, Vittoria Cammisotto, Cammisotto Nocella et al. "Extra Virgin Olive Oil Reduces Gut Permeability and Metabolic Endotoxemia in Diabetic Patients." *Nutrients* 14, no. 10 (2022). https://doi.org/10.3390/nu14102153.

Bozzetto, Lutgarda, Antonio Alderisio, Gennaro Clemente et al. "Gastrointestinal Effects of Extra-Virgin Olive Oil Associated with Lower Postprandial Glycemia in Type 1 Diabetes." *Clinical Nutrition* 38, no. 6 (2019): 2645–51. https://doi.org/10.1016/j.clnu.2018.11.015.

Carnevale, Roberto, Lorenzo Loffredo, Maria Del Ben et al. "Extra Virgin Olive Oil Improves Post-Prandial Glycemic and Lipid Profile in Patients with Impaired Fasting Glucose." *Clinical Nutrition* 36, no. 3 (2017): 782–87. https://doi.org/10.1016/j.clnu.2016.05.016.

Luisi, Maria Luisa Elianna, Laura Lucarini, Barbara Biffi et al. "Effect of Mediterranean Diet Enriched in High Quality Extra Virgin Olive Oil on Oxidative Stress, Inflammation and Gut Microbiota in Obese and Normal Weight Adult Subjects." *Frontiers in Pharmacology* 10 (2019). https://doi.org/10.3389/fphar.2019.01366.

Morvaridzadeh, Mojgan, Alan A. Cohen, Javad Heshmati et al. "Effect of Extra Virgin Olive Oil on Anthropometric Indices, Inflammatory and Cardiometabolic Markers: A Systematic Review and Meta-Analysis of Randomized Clinical Trials." *Journal of Nutrition* 154, no. 1 (2023): 95–120. https://doi.org/10.1016/j.tjnut.2023.10.028.

Violi, Francesco, Lorenzo Loffredo, Pasquale Pignatelli et al. "Extra Virgin Olive Oil Use Is Associated with Improved Post-Prandial Blood Glucose and LDL Cholesterol in Healthy Subjects." *Nutrition & Diabetes* 5, no. 7 (2015): e172. https://doi.org/10.1038/nutd.2015.23.

Broccoli/Broccoli Sprouts

Bouranis, John A., Laura M. Beaver, Carmen P. Wong et al. "Sulforaphane and Sulforaphane-Nitrile Metabolism in Humans Following Broccoli Sprout Consumption: Inter-individual Variation, Association with Gut Microbiome Composition, and Differential Bioactivity." *Molecular Nutrition & Food Research* 68, no. 4 (2024): e2300286. https://doi.org/10.1002/mnfr.202300286.

Mirmiran, Parvin, Zahra Bahadoran, Farhad Hosseinpanah, Amitis Keyzad, and Fereidoun Azizi. "Effects of Broccoli Sprout with High Sulforaphane Concentration on Inflammatory Markers in Type 2 Diabetic Patients: A Randomized Double-Blind Placebo-Controlled Clinical Trial." *Journal of Functional Foods* 4, no. 4 (2012): 837–41. https://doi.org/10.1016/J .JFF.2012.05.012.

Yanaka, Akinori. "Daily Intake of Sulforaphane-Rich Broccoli Sprouts Normalizes Bowel Habits in Healthy Human Subjects." *FASEB Journal* 31, no. S1 (2017). https://doi.org/10.1096/fasebj.31.1_supplement.972.20.

Cacao/Cocoa

Grassi, Davide, Giovambattista Desideri, Stefano Necozione et al. "Blood Pressure Is Reduced and Insulin Sensitivity Increased in Glucose-Intolerant, Hypertensive Subjects After 15 Days of Consuming High-Polyphenol Dark Chocolate." *Journal of Nutrition* 138, no. 9 (2008): 1671–76. https://doi .org/10.1093/JN/138.9.1671.

Lin, Xiaochen, Isabel Zhang, Alina Li et al. "Cocoa Flavanol Intake and Biomarkers for Cardiometabolic Health: A Systematic Review and Meta-Analysis of Randomized Controlled Trials." *Journal of Nutrition* 146, no. 11 (2016): 2325–33. https://doi.org/10.3945/jn.116.237644.

Petyaev, Ivan M., and Yuriy K. Bashmakov. "Cocobiota: Implications for Human Health." *Journal of Nutrition and Metabolism* 2016, no. 8 (2016): 1–3. https://doi.org/10.1155/2016/7906927.

Strat, Karen M., Thomas Rowley, Andrew T. Smithson et al. "Mechanisms by Which Cocoa Flavanols Improve Metabolic Syndrome and Related Disorders." *Journal of Nutritional Biochemistry* 35, no. S10 (2016): 1–21. https:// doi.org/10.1016/j.jnutbio.2015.12.008.

SWITCH OPERATIONS AND IMPACT

Arch G. Mainous, Lu Yin, Velyn Wu et al. "Body Mass Index vs Body Fat Percentage as a Predictor of Mortality in Adults Aged 20-49 Years." *Annals of Family Medicine* 23, no. 3 (June 2025). https://doi.org/10.1370/afm.240330.

Baggio, Laurie L., and Daniel J. Drucker. "Biology of Incretins: GLP-1 and GIP." *Gastroenterology* 132, no. 6 (2007): 2131–57. https://doi.org/10.1053/j .gastro.2007.03.054.

Bilchik, Anton J., Oscar Joe Hines, Thomas E. Adrian et al. "Peptide YY Is a Physiological Regulator of Water and Electrolyte Absorption in the Canine Small Bowel In Vivo." *Gastroenterology* 105, no. 5 (1993): 1441–48. https:// doi.org/10.1016/0016-5085(93)90149-7.

Bodur M., and R. Nerqiz Unal. "The Effects of Dietary High Fructose and Saturated Fatty Acids on Chronic Low Grade Inflammation in the Perspective of Chronic Diseases." *Cukurova Medical Journal* 44, no. 2 (2019): 685–94. https://doi.org/10.17826/cumj.482623.

Cabou, Cendrine, and Rémy Burcelin. "GLP-1, the Gut-Brain, and Brain-Periphery Axes." *Review of Diabetic Studies* 8, no. 3 (Fall 2011): 418–31. https://doi.org/10.1900/RDS.2011.8.418.

Jensterle, Mojca, Simona Ferjan, Tadej Battelino et al. "Does Intervention with GLP-1 Receptor Agonist Semaglutide Modulate Perception of Sweet Taste in Women with Obesity: Study Protocol of a Randomized, Single-Blinded, Placebo-Controlled Clinical Trial." *Trials* 22 (2021). https://doi .org/10.1186/s13063-021-05442-y.

Jones, Lauren A., Emily W. Sun, Amanda L. Lumsden et al. "Alterations in GLP-1 and PYY Release with Aging and Body Mass in the Human Gut." *Molecular and Cellular Endocrinology* 578 (2023). https://doi.org/10.1016/j .mce.2023.112072.

Ko, Byeong-Seong, Joung-Ho Han, Jee-In Jeong et al. "Mechanism of Action of Cholecystokinin on Colonic Motility in Isolated, Vascularly Perfused Rat Colon." *Journal of Neurogastroenterology and Motility* 17, no. 1 (2011): 73–81. https://doi.org/10.5056/jnm.2011.17.1.73.

Martin, Bronwen, Cedrick D. Dotson, Yu-Kyong Shin et al. "Modulation of Taste Sensitivity by GLP-1 Signaling in Taste Buds." *Annals of the New York Academy of Sciences* 1170, no. 1 (2009): 98–101. https://doi.org/10.1111 /j.1749-6632.2009.03920.x.

Shin Yu-Kyong, Bronwen Martin, Erin Golden et al. "Modulation of Taste Sensitivity by GLP-1 Signaling." *Journal of Neurochemistry* 106, no. 1 (2008): 455–63. https://doi.org/10.1111/j.1471-4159.2008.05397.x.

SWITCH OPTIMIZATION LIFESTYLE

Ataeinosrat, Ali, Marjan Mosalman Haghighi, Hossein Abednatanzi et al. "Effects of Three Different Modes of Resistance Training on Appetite Hormones in Males with Obesity." *Frontiers in Physiology* 13 (2022). https://doi .org/10.3389/fphys.2022.827335.

Phillips-Wren, Gloria, and Monica Adya. "Decision Making Under Stress: The Role of Information Overload, Time Pressure, Complexity, and Uncertainty." *Journal of Decision Systems* 29, no. S1 (2020): 213–25. https://doi.org /10.1080/12460125.2020.1768680.

Roenneberg, Till. "How Can Social Jetlag Affect Health?" *Nature Reviews Endocrinology* 19, no. 7 (2023): 383–84. https://doi.org/10.1038/s41574 -023-00851-2.

Rood, Kara M., Niharika Patel, Ivana M. DeVengencie et al. "Aspirin Modulates Production of Pro-Inflammatory and Pro-Resolving Mediators in Endothelial Cells. *PLoS One* 18, no. 4 (2023): e0283163. https://doi .org/10.1371/journal.pone.0283163.